Brain Development and Sexual Orientation

Colloquium Series on
The Developing Brain

Editor
Margaret M. McCarthy, PhD,
Professor and Chair
Department of Pharmacology
University of Maryland School of Medicine

The goal of this series is to provide a comprehensive state-of-the-art overview of how the brain develops and those processes that affect it. Topics range from the fundamentals of axonal guidance and synaptogenesis prenatally to the influence of hormones, sex, stress, maternal care, and injury during the early postnatal period to an additional critical period at puberty. Easily accessible expert reviews combine analyses of detailed cellular mechanisms with interpretations of significance and broader impact of the topic area on the field of neuroscience and the understanding of brain and behavior.

My research program focuses on the influence of steroid hormones on the developing brain. During perinatal life, there is a sensitive period for hormone exposure during which permanent cytoarchitectural changes are established. Males and females are exposed to different hormonal milieus and this results in sex differences in the brain. These differences include alterations in the volumes of particular brain nuclei and patterns of synaptic connectivity. The mechanisms by which sexually dimorphic structures are formed in the brain remains poorly understood.

I received my PhD in Behavioral and Neural Sciences from the Institute of Animal Behavior at Rutgers University in Newark, NJ in 1989. I then spent three years as a post-doctoral fellow at the Rockefeller University in New York, NY and one year as a National Research Council Fellow at the National Institutes of Health, before joining the faculty at the University of Maryland. I am a member of the University of Maryland Graduate School and the Center for Studies in Reproduction. I am also a member of the Society for Behavioral Neuroendocrinology, the Society for Neuroscience, the American Physiological Association, and the Endocrine Society.

Brain Development and Sexual Orientation
Jacques Balthazart
www.morganclaypool.com

ISBN: 9781615044580   paperback

ISBN: 9781615044597   ebook

DOI: 10.2200/C00064ED1V01Y201208DBR008

A Publication in the

COLLOQUIUM SERIES ON THE DEVELOPING BRAIN

Lecture #8

Series Editor: Margaret M. McCarthy, University of Maryland School of Medicine

Series ISSN
ISSN 2159-5194   print
ISSN 2159-5208   online

Brain Development and Sexual Orientation

Jacques Balthazart
University of Liège

COLLOQUIUM SERIES ON THE DEVELOPING BRAIN #8

ABSTRACT

Sexual orientation (homo- vs. heterosexuality) is one of many sex differences observed in humans. Sex differences can result from differential postnatal experiences (interaction with parents, environment) or from biological factors (hormones and genes) acting pre- or postnatally. The first option is often favored to explain sexual orientation although it is supported by little experimental evidence. In contrast, many sexually differentiated behaviors are organized during early life by an irreversible action of sex steroids. In particular, the preference for a male or female sex partner is largely determined in rodents by embryonic exposure to sex steroids. The early action of these steroids also seems to affect sexual orientation in humans. Indeed, clinical conditions associated with major endocrine changes during embryonic life often result in an increased incidence of homosexuality. Furthermore, multiple sexually differentiated behavioral, physiological, or even morphological traits that are known to be organized by prenatal steroids, at least in animals, are significantly different in homo- and heterosexual populations. Thus, prenatal endocrine (or genetic) factors seem to influence significantly human sexual orientation even if a large fraction of the variance remains unexplained to date. The possible interaction between biological factors acting prenatally and postnatal social influences remains to be investigated.

KEYWORDS

sexual differentiation, sexual orientation, homosexuality, estrogens, androgens, preoptic area, congenital adrenal hyperplasia, sexually dimorphic nucleus, testosterone

Contents

Contents

CHAPTER 1

Introduction and Definitions

1.1 BACKGROUND AND GOALS OF THIS PRESENTATION

The last few decades have seen an amazingly fast development of biochemical, molecular, anatomical and imaging techniques that have led to an equally impressive accumulation of knowledge concerning mechanisms that control animal behavior. This field of research has now reached a level of maturity that allows asking whether specific mechanisms identified in animal studies are also relevant in humans. One of these questions concerns the determination of sexual orientation. Depending on the society considered, a large fraction of the population ranging from about 40% up to 90% is convinced that sexual orientation (homosexuality or heterosexuality) is either the result of a deliberate choice of adults or the consequence of events, in particular interaction with parents, educators and siblings that took place during early childhood. These notions were originally proposed by Sigmund Freud (Freud, 1905/1975) (see LeVay & Valente, 2006 for summary) and have been perpetuated and expanded by his successors, but they have surprisingly received very little pragmatic test and even less experimental confirmation.

In contrast, it is well established in animals that many sexually differentiated behavioral characteristics are organized during the embryonic or early postnatal life by an irreversible action of sex steroids. It has in particular been demonstrated that the preference for a male or female sex partner is largely determined in rodents by the embryonic exposure to the sex steroids testosterone and its metabolite estradiol. Many researchers have thus wondered whether the same mechanism could also affect human sexual orientation.

Two types of evidence actually support this notion even if its formal demonstration remains impossible essentially for ethical and pragmatic reasons. First, clinical conditions associated with significant endocrine changes during embryonic life often result in an increased incidence of homosexuality. Second, multiple sexually differentiated behavioral, physiological or even morphological traits are significantly different in homo- and heterosexual populations. Because some of these traits are known to be organized by prenatal steroids, these differences suggest that homosexual subjects were, on average, exposed to atypical endocrine conditions during their embryonic and early postnatal development. Genetic differences affecting behavior either in a direct manner or by changing embryonic hormones secretion or action also seem to be involved. Thus, prenatal endocrine or genetic factors seem to have a significant influence of human sexual orientation even if a large fraction of the variance in this behavioral characteristic remains unexplained to date. In particular, the possible interaction between these biological factors acting prenatally and the postnatal social influences remains to be investigated.

I shall, in this book, review the evidence supporting the idea that human sexual orientation, and possibly also gender identity, is controlled by prenatal factors of endocrine or genetic nature. Specifically, I shall first explain how sexual orientation and gender identify can be considered as some of the most differentiated behavioral characteristics in humans. Then, I shall describe the research literature on animals demonstrating how behavioral characteristics, sexual partner preference in particular, are markedly affected by the very early organizing action of sex steroids. One chapter will then review the evidence that clinical conditions resulting in an atypical endocrine environment during fetal life are regularly associated with an increased incidence of homosexuality in men and women. The next two chapters will summarize data coming from a large number of studies indicating that multiple morphological, physiological or behavioral characteristics that bear no direct relationship with reproduction are different in homosexual and heterosexual populations. Because these traits are almost always sexually differentiated (different in men and women) and are often known to develop under the influence of sex steroids during early life (a fact largely established by experiments in animals and often strongly suspected in humans), these atypical traits suggest that individuals presenting a homosexual orientation in adulthood were exposed in early life to abnormal endocrine conditions for their sex. I shall then summarize another set of studies taking a completely independent although not contradictory approach indicating that genetic factors also contribute significantly to the determination of sexual orientation. In the last chapter, I shall finally ask to what extent postnatal social factors interact with the prenatal factors described in this book to control sexual orientation.

However, before embarking for this trip into mechanisms of neuroendocrine and genetic control, it is first necessary to provide a few essential definitions in order to avoid misunderstandings that are in this field unfortunately much too frequent.

1.2   SEXUAL MOTIVATION, SEXUAL ORIENTATION, GENDER IDENTITY AND GENDER ROLE

Sexual orientation is only one of the many dimensions of human sexuality. These different features of human sexual behavior are often confounded or at least referred to with improper if not inaccurate terms. The notion of sexual orientation (homosexuality vs. heterosexuality) is also complex in itself and reflects several aspects of behavior or cognition including attraction towards a given sex, performance of sexual acts with that partner and finally acceptance of this sexual orientation and its public declaration. Because these notions are often poorly understood, it is critical to define them first, in order to clarify what is really the topic of this presentation.

Sexologists distinguish at least five more or less independent dimensions of human sexuality: 1) the specific action patterns that are produced by the individual (erections, pelvic thrusts, orgasm, . . .) sometimes grouped under the term performance, 2) the motivation underlying expression of these behaviors, 3) the orientation of these behaviors and of the associated sexual fantasies toward a partner of the same or of the opposite sex (sexual orientation), 4) the sex that the individual believes he or she has and is usually called sexual or gender identity, and finally 5) the sexual role that the individual plays in the society. These different aspects of sexuality are usually correlated and all these characteristics can usually be defined as being either male or female within an individual. However, discrepancies do occur and then require a specific vocabulary to precisely describe the behavioral or cognitive features being observed.

Sexual motivation, orientation, identity and role are linked to and, as this presentation should illustrate, largely determined by the genetic and gonadal (testes vs. ovaries) sex. When dissociations are observed between these characteristics, which happens in a small but significant fraction (a few percents) of the population, scientists (sexologists) speak of **homosexuality** when the sexual orientation of an individual does not correspond to his or her morphological (and in general genetic) sex. **Transsexuality** refers to the gender identity of an individual who will be considered transsexual if his or her identity is at odds with the morphological sex. Finally **transvestism** is invoked when the gender role played by an individual in society does not correspond to his/her morphological sex.

The first two aspects of human sexuality, specifically, the **performance** of sexual acts and the **motivation** supporting them, have clear equivalent in animals and can thus be studied experimentally in a variety of model species. The behavior patterns displayed by most animals are sexually differentiated. In rodents, for example, males show mounts, pelvic thrusts, intromissions culminating in ejaculation, whereas females display a variety of proceptive behaviors and a receptive behavior called lordosis. These behaviors are controlled by steroid hormones both during development (organization; see McCarthy & Ball, 2008; McCarthy, 2011) and during adulthood (activation). In contrast, human sexual behavior is by and large not limited to sexually differentiated motor acts. Instead of having stereotyped sex-specific mating postures, as is the case in rats (mount and intromission in the male, lordosis in the female), the human species mates in a variety of positions. A position is often preferred in a given culture (e.g., man over the woman or "missionary position" in Western civilizations), but other positions are commonly used (Nelson, 2011). Men and women do not play stereotypical roles in the sexual act, there is no major difference between the motor acts performed by the two sexes. Consequently, the animal literature on mechanisms that control sex differences affecting this aspect of behavior has unfortunately only an indirect interest for understanding the human species.

The motivation that leads human beings to have sex is, like in animals, very high in both sexes. This feature of human and animal sexuality is therefore not affected by major sex differences.

Differences in modality have been reported, but they are minor and often related to culture. This motivation is in humans, as it is in animals, largely under the control of sex steroids.

If the motor aspects of sexual behavior and its motivation are quite similar among men and women, the three other dimensions of sexuality are highly differentiated. **Sexual orientation** is defined by the sex of person(s) to which an individual directs his/her behavior but also his/her sexual fantasies. Most men and women are sexually attracted and excited by individuals of the opposite sex. They are heterosexual. However, there is a reliable percentage of subjects who are attracted to persons of their sex, they are homosexuals. This distinction is not necessarily qualitative. Kinsey and colleagues in the late 1940s already recognized that all intermediaries could exist (Kinsey et al., 1948). It is considered that 3% to 10% of men are homosexuals in all cultures. These numbers vary slightly, depending on the country and study methods but are surprisingly stable within this range, irrespective of the social attitude toward homosexuals.

The estimates for women are more variable, but probably in the same order of magnitude or slightly higher with the addition of a significant population of individuals regarding themselves as bi-sexual (attraction to women as well as for men) (Kinsey et al., 1953). There is thus in all human populations a significant proportion of gays and lesbians, but this percentage is usually smaller than 10%, so that the majority of the population is heterosexual. Sexual orientation therefore represents a sexually differentiated behavioral traits since most men and women differ with regard to this trait.

Regardless of sex of the partner found interesting or exciting from a sexual point of view, each human being is also confident of belonging to one sex, either male or female. This conviction is unchangeable, and it often seems to develop early in infancy. Most people think that they have a **sexual identity** that is in agreement with their genetic sex and their genital morphology. However, a small number of men and women are convinced to the contrary and believe that they were "born in a body that does not correspond to their sex." They are called transsexuals. This atypical gender identity develops early in infancy (usually before 3 years of age) and contrary to what was thought until the last parts of the 20th century, it is usually difficult, if not impossible, to change this sexual identity later (see also the History of John/Joan, Chapter 13). This observation has led to the speculation that gender identity is actually the consequence of endocrine events that take place during embryonic life (Swaab, 2007) and clinical data support this interpretation (see Chapter 9). This idea is difficult to assess because experiments on this topic are impossible in humans for obvious ethical reasons and there is no animal model for studying sexual identity. It is indeed impossible to ask an animal what sex it belongs assuming that it would be aware of its own body and sex. We will nevertheless discuss in Chapters 9 and 10 the limited clinical data pointing to the idea that sexual identity is at least partly determined by the embryonic hormonal conditions.

Finally, a last dimension of sexuality refers to the role played by an individual in society and is called **gender role**. Sex differences affecting this role are deeply influenced by the structure of the society in which the individual lives at a given time. Thus, housework tasks were typically performed by women in Western Europe until the middle of the last century, but are now widely (perhaps not quite enough yet!) shared by men and women. Agriculture is typically a male or female task depending on the human group considered. It would probably be futile to search for significant biological bases for this type of sex difference, even if sociobiology suggests that selective pressures acting on our ancestors may have selectively adjusted men and women for various jobs (but this idea is broadly contested).

As already mentioned, sex differences concerning these various dimensions of sexuality are usually correlated and in agreement with each other in the majority of individuals. A genetic male is usually attracted by women and has a masculine gender identity. This correlation can be expected both on the basis of coordinated learning that affects these different features as well as on the basis of a prenatal hormonal milieu that plays a role in their organization. They must however be clearly distinguished because they are exceptionally discordant and it is then, in particular, critical to distinguish between sexual orientation and gender identity. A male (female) subject may be homosexual but still consider him/herself as a man (a woman) and thus have a male (female) sexual identity. Additionally, a transsexual man who believes to be a woman (male-to-woman transsexual in the literature) can be sexually attracted by men or by women. Classifying this orientation is then tricky and will depend on the sex of reference that is considered (genetic and genital sex or sex defined by gender identity).

1.3 THE DIFFERENT FACETS OF SEXUAL ORIENTATION

Two terms are commonly used to describe the attraction for sexual partners of one sex or the other: sexual orientation or sexual preference. These terms clearly have different connotations. Sexual orientation is commonly used by people who believe that the homo- or heterosexuality is due mainly to biological factors while sexual preference rather suggests that the homo- or heterosexuality is a lifestyle that is learned or chosen in a more or less deliberate fashion during life. In my view, orientation is also more neutral and will be used almost exclusively.

It is also critical to clearly define what will be discussed in this book. Homosexuality can indeed refer to a) the attraction to sexual partners of the same sex, b) the expression of sexual behaviors with these partners and finally c) the acceptance and public recognition of this attraction. The discussion of the biological bases of homosexuality that will follow will be concerned only with the first of these aspects, the sexual attraction to partners of the same sex irrespective of whether it is publicly expressed, and it actually leads to homosexual relationships. These last two aspects of homosexuality are indeed largely conditioned by the (social) environment. For example, homosexual behavior occasionally takes place in circumstances where partners of the opposite sex are not available (e.g., prisons or single-sex teaching institutions) but when they are available again, most subjects will return to heterosexual sex. There is no preferential homosexual attraction in these cases. The homosexual activity is simply an outlet for sexual motivation. Similarly, the personal acceptance and public disclosure (coming out) of homosexual orientation is by and large a conscious act that largely depends on the willingness of the person and the social context. It is thus also controlled mostly by the environment, not so much the biology.

In contrast sexual attraction is not or only marginally conditioned by the environment and reflects an internal aspiration that predates the first sexual relationships. It is the preferential or exclusive attraction to individuals of the same sex that we are talking about mainly in this book. We will show that this attraction depends, probably to a large extent, on prenatal biological phenomena that are largely beyond the will of the concerned individual. Note, however, that homosexuality and sexual orientation, in general, is a complex phenomenon that is probably not the result of a single cause. In addition, male and female homosexuality may have different explanations, at least in part. We will come back to this idea at the end of this book when discussing the potential contribution of social influences and education or conscious choices made in adulthood.

Multiple causes always interact to control human behavior. They include the structure of the genome and its individual variations, the developmental history, including the social, hormonal and environmental stimuli to which the individual has been exposed during its life and its physiological condition. In most cases, it is thus impossible to identify a single specific cause to a behavior. A given factor will only contribute to the explanation of a behavior for a certain percentage of the observed variance. This does not, however, invalidate the deterministic nature of behavior.

• • • •

CHAPTER 2

Sex Differences in Animals and Men

Men and women differ in their morphology and physiology. No one would ever dispute the fact that on average, men are taller, have more facial hair and have a larger muscular mass than women, whereas women are characterized by the presence of breast and larger hips. Multiple physiological differences between the sexes have also been identified in humans (Hines, 2010; McCarthy et al., 2012). For example, liver metabolism is very differentiated between men and women, with the latter metabolizing more slowly a number of compounds such as alcohol. These differences thus extend outside the field of reproduction in its narrower sense. In the reproductive realm, differences obviously abound. They concern the morphology of the internal and external sex organs as well as the physiology of the gonads (testes in men and ovaries in women) and of the pituitary gland that controls gonadal activity.

When it comes to brain and behavior, however, skepticism is more widespread if one tries to assert that men and women might be different (Vidal, 2007; Jordan-Young, 2010; Fine, 2011). If we first consider the brain, it was thought for a long time that the brains of males and females (men and women) were identical if one compensates for the usually larger body size in males. However, in 1976, large sex differences were identified in the volume of brain regions that control singing behavior in birds such as canaries and zebra finches (Nottebohm & Arnold, 1976), and this led other researchers to re-consider the notion that male and female brains are identical. In 1978, Roger Gorski and his team at the University of California, Los Angeles described in the preoptic area of rats a small nucleus that was approximately five times larger in males than in females (Gorski et al., 1978). This was the beginning of a very active line of research that has demonstrated over the next decades the presence of multiple sex differences in the brain of many mammalian species including humans.

Now that the visualization and quantification of the volume of brain parts has become possible by imaging techniques in live subjects, it has even become clear that most parts of the human brain have a significantly different volume in men and women. Sex differences in brain morphology are thus no longer the exception, they tend to be the rule (Cahill, 2006).

Additionally, it has now been firmly established that the incidence of most diseases affecting the nervous system is different in men and women. Men are more frequently affected by autism (male/female incidence, 5:1) or schizophrenia (ratio 2:1) while women are more often affected by depression (female/male, 2:1), Alzheimer's disease (2:1) or anorexia nervosa (4:1) (Cahill, 2006;

Becker et al., 2008). Dyslexia is also twice as common among boys than girls in both Europe and USA.

For behavior and cognitive skills, resistance to the potential existence of sex differences is even stronger. This relates in part to legitimate concerns about the request for equal rights and equal access to various professions of both sexes but also partly to unsupported resistance to the idea that the two sexes might be unequal, even more so when it is suggested that the difference might have a biological origin.

Behavioral differences between men and women are however plentiful. Men are on average significantly more aggressive than women. There are also cognitive differences between sexes. Many studies have identified the existence of reproducible differences in verbal skills (use of language and ability to spell words: F > M) that favor women over men, but on the other hand, men have better skills in certain aspects of mathematical reasoning and analysis of spatial relationships between objects (see (Ellis et al., 2008) for a relatively complete catalog of these innumerable differences).

The existence of such differences is not surprising when considering that in animals (and probably also in humans) the majority of genes are differentially expressed in males and females. A recent study showed that in mice, 72% of genes in the liver and 14% of genes expressed in the brain are differentially expressed in males and females (Yang et al., 2006). One should probably consider the existence of differences between sexes as the rule rather than the exception even if the magnitude of these differences can vary very much depending on the species and the character in question. Epidemiological or pharmacological studies would do well to take such differences into account more systematically, because many experimental results demonstrated in a population of one sex may well not apply directly to the other sex (Blaustein, 2012; McCarthy et al., 2012). A dose of medication that is effective and safe in one sex may be less effective or have side effects in the other sex (see the website of the Organization for the Study of Sex Differences, or OSSD at: http://www .ossdweb.org/). The burden of the proof has thus shifted camp: one can consider that it is the lack of difference between males and females that must now be demonstrated not the presence of such a difference. The difference can as a rule always be assumed by default.

In summary, differences between men and women affecting morphology, physiology and behavior are very abundant and, in fact, far too numerous to be summarized here. A very detailed book of 900 pages was published on this issue including references to more than 22,000 scientific studies on the subject (Ellis et al., 2008). The magnitude of these differences can however be extremely variable, and this notion is often poorly understood. A short presentation of how scientists apprehend the magnitude of a difference is necessary here to clarify the presentation.

2.1 THE NOTION OF EFFECT SIZE

As illustrated in Figure 1, a sex difference can be statistically significant and highly reproducible even if some overlap exists between the sexes in the measure of this variable. Such an overlap can

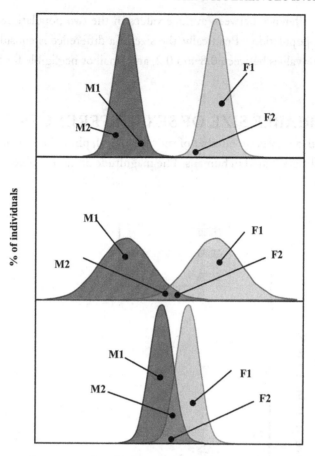

FIGURE 1: Variability in the magnitude of differences between two populations, in this case, males (in blue) and females (in pink). In all three panels, female scores are on average larger than male scores. In the top panel, there is no overlap between scores in the two sexes so that **all** females have a higher score than males. A reliable and statistically significant different can however still be present when the scores of both sexes partly overlap either because variance is larger within one sex (middle panel) or difference between means is smaller (bottom panel). In these cases, the score of male 2 (M2) can be equal or even larger than the score of female 2 (F2). It is therefore impossible to predict the sex of a subject based on his/her score but this does not mean, however, that there is no statistically significant difference between males and females.

occur because either the average difference between sexes is small or the variability within one sex is large. Scientists have developed an integrated measure that takes these two aspects of the differences into account to provide a single measure that will accurately depict the degree of overlap between two populations. This measure is called the effect size.

The effect size (d) is directly proportional to the difference between the means of the two populations but inversely related to the variance in these populations. For example, an effect size of 2 means that the difference between average values in the two populations is twice as large as the variance in these populations. Practically, the size of a difference is considered large if greater than 0.8, moderate for values between 0.8 and 0.2, and small or negligible for lower values (Hines, 2004).

2.2   THE VARIABLE SIZE OF SEX DIFFERENCES

As already mentioned, a very large number of morphological, physiological and behavioral charac-teristics are sexually differentiated in humans. The magnitude of these differences is however highly

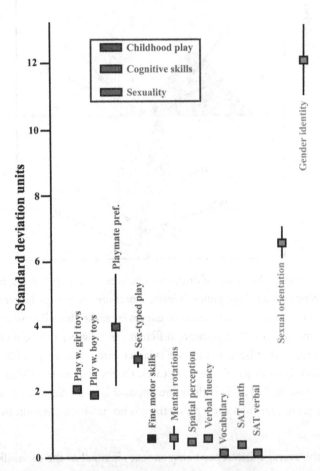

FIGURE 2: Magnitude of sex differences affecting various human characteristics. The amplitude is measured by the effect size (difference between means divided by the variance of the population). The effect size is large (>0.8) for childhood play but generally low for behavioral or cognitive differences. Sexual orientation and gender identity constitute the largest sex differences identified so far in humans. Drawn from data in Hines (2010).

variable. The effect size of sex differences for many morphological variables is large (>0.8). This is true for measures such as the height of an individual and also for neuroanatomical differences such as those affecting the volume of the dimorphic nucleus of the preoptic area. This nucleus is indeed about twice larger in men than in women (Swaab & Fliers, 1985). In contrast, at the behavioral level, the size of differences in statistical terms is often small to negligible (difference between means equal to less than 20% of the common variance of data; i.e., d = 0.2). For example, the differences in cognitive abilities (e.g., overall verbal capacity) have an effect size that is low to negligible (<0.2), but there are exceptions. Thus, mental rotation skills of three-dimensional objects are associated with a sex difference that has a size considered large (Figure 2).

The small magnitude of the sex differences in cognitive abilities clearly means that in everyday life, there is no reason to establish any discrimination between the sexes. Even if a job requires a skill known to be sexually differentiated, and for which men are on average more efficient, it is very likely that women candidates will be found for the position that are better than male candidates (and vice versa). Any discrimination based on sex is thus not justified by the existence of the statistical differences of small magnitude that we are talking about here.

That being said, it is also not reasonable to deny the existence of these differences on the basis of *a priori* preconceptions usually related to egalitarian theories. A difference does not imply superiority. The fight against discrimination is laudable, but will be more effective if it is based on a correct analysis of the real situation. There are obvious behavioral differences between men and women. The real question is their origin, biological (genetic, hormonal) or environmental (education, social)? We will come back to this question in the next chapter.

2.3 SEXUAL ORIENTATION AND SEXUAL IDENTITY ARE THE MOST DIFFERENTIATED CHARACTERISTICS IN HUMANS

Sexual orientation can be considered as a sexually differentiated characteristic in humans since most men display a sexual attraction (toward women) that is different from the attraction shown by women. Thus, in the vast majority of the population (well over 90%), there is no overlap between the sexual orientation of men and women. The effect size associated with such a large difference is extremely large ($d = 6$), much larger than the difference in height ($d = 2$) or in verbal fluency ($d = 0.3$) (Hines, 2004; 2010). Homosexuality in this context represents the absence of difference between sexes in terms of sexual orientation. A gay man has the same sexual attraction as a heterosexual woman, whereas a homosexual woman shows the same preference as heterosexual males. Note that I only consider here the sex of the individual who will be found sexually attractive; it is clear that gay men and women will not be attracted by exactly the same type(s) of men.

Incidentally, it should also be noted here that the sex difference related to gender identity has an even larger magnitude ($d = 11$) (Hines, 2010). Although it is very difficult to obtain accurate estimates of the number of people who believe that they belong to the other sex (as compared to their genetic and morphological sex), the incidence of transsexualism is estimated to range between 1 to 20,000 and 1 to 100,000 subjects, meaning that over 99.9% of individuals in one sex differ from subjects of the other sex in terms of their sexual identity. U.S. statistics estimate that 1 in 30,000 men and 1 woman in 100,000 seek medical treatment and surgery to change sex. Less important disorders of sexual identity, including the desire to be of the opposite sex, but falling short of demanding surgery, are probably more frequent, but accurate statistics are difficult to obtain. Statistics from the Netherlands, where medical and psychological help is widely available for patients suffering from sexual identity problems, indicates the presence of these disorders in about one man out of 20,000 and one woman in 50,000 (Hines, 2004).

2.4   OCCURRENCE IN THE WORLD

Homosexuality is clearly observed in all Western societies and is probably present in all extant societies as it was in the past. Because more or less negative or even hostile reactions are present in most societies, the prevalence of homosexuality and bisexuality has always been difficult to assess. The first attempts of objective quantification in men and women were conducted and published by the zoologist Alfred Kinsey and his collaborators in the mid-twentieth century (Kinsey et al., 1948; Kinsey et al., 1953). Kinsey concluded that about 10% of men are exclusively or nearly exclusively homosexual (Kinsey et al., 1948), while the corresponding proportion was about 1.5% among women (Kinsey et al., 1953).

More recent studies based on approaches ensuring an absolute anonymity of respondents led to slightly different usually lower estimates. A U.S. study organized by the National Health Statistics Center of the Center for Disease Control (NHSC) indicates, for example, that 7.1% of men and 13.6% of women feel sexual attraction for people of their own sex (Mosher et al., 2005). But only 1.5% of men and 0.7% of women reported exclusive homosexual attraction (see Figure 3). These data could be an underestimate, and it has been claimed that this is indeed the case, but it is likely that they are closer to reality than the estimates of Kinsey and collaborators given the precautions taken to ensure as much as possible of the objectivity of the responses of subjects and the representativeness of the analyzed samples.

Despite the biases that are potentially present in all such studies, it is interesting to note that the reported percentages of homosexuality in the United States are in the same range as the percentages reported in anthropological studies of various societies, independently of whether these societies have been influenced much by the Western world and independently of their very permissive or repressive attitude towards homosexuality. In all societies studied recently or in the past, this percentage varies between 2–3% and 10% (see Table 2.1).

It should also be noted that although some variation is observed, this percentage is relatively similar in the study of Kinsey and of the NHSC performed more than 50 years apart. Since the at-

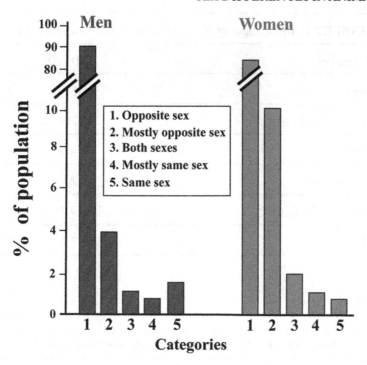

FIGURE 3: Distribution of sexual attraction in the U.S. population of men and women aged 18 to 44 years (according to data from Mosher et al., 2005).

titude of society towards homosexuals has profoundly changed during this period, one could expect an increase in their relative numbers if a permissive environment had an influence on sexual orientation. But, if anything it is rather the opposite that has been observed for men.

As shown in Figure 3, the vast majority of individuals of both sexes have a preferential or exclusive attraction to individuals of the opposite sex. Interestingly however, the curves describing the distribution of sexual orientation are different for men and women. In men, there is clearly a bimodal distribution that has two peaks at the extremes (strictly hetero and strictly homosexual), while among women there is rather a gradual decrease of percentages when moving from an exclusively hetero- to an exclusively homosexual orientation. Thus, the categories homo- vs. heterosexual people describe relatively discrete populations in men while this distinction is more subtle in women.

In women, bisexuality, defined as the existence of any degree of attraction for both sexes, seems to be more common than strict homosexuality and is definitely more common in women than in men. Some sexologists even contest that such a bisexuality is present in men at all. Measures

TABLE 2.1: Percentages of homosexual men in different societies.	
Great Britain	5.0–9.0
Japan	5.8
The Netherlands	7.8
USA	4.8
Philippines*	2.0
Pilau (Malaysia)*	4.7
Thailand*	3.6
Mean	**4.8–5.4**

The table lists average percentage of male homosexuality in several European, North American and Asian countries that differ widely in their attitude toward homosexual orientation. The 3 countries whose name is followed by an asterisk are very tolerant and even favor adolescent male homosexuality. This does not seem to affect the occurrence of homosexual orientation in adulthood (data from Diamond, 1993).

of genital arousal (tumescence and erection) in men who were presented erotic pictures of men and women have shown that men who call themselves bisexual show genital arousal much more and often exclusively for men pictures (Freund, 1974; Rieger et al., 2005). Levay and Valente note that the term bisexual is frequently used by young men who are in the process of becoming aware and publicly revealing their homosexual orientation ("do their coming-out") (LeVay & Valente, 2006). Some studies show that a percentage of homosexuals that can be as high as 40% identified themselves at an early stage of their young adult life as a bisexual before describing themselves later as homosexual. As Michael Bailey puts it in his popular book: "A man is either heterosexual or homosexual or is a liar!" (Bailey, 2003).

Female bisexuality is in contrast more widely recognized. At the genital level, most women are in fact bisexual and show reactions of genital arousal in response to erotic videos showing both men and women regardless of their declared attraction to one sex or the other (Chivers et al., 2004). These data do not, however, exclude the existence of exclusive homo-or heterosexuality among women. They only show that the physiological genital arousal can be (partly) independent of verbal statements of sexual attraction.

Correspondingly, the stability of sexual orientation is greater in men than in women. Longitudinal studies show that bisexual women or women with a non-exclusive preference for one sex change more frequently their sexual orientation in the course of their lifetime than men. Sexual orientation is however generally stable over time in both sexes. Only a small percentage of subjects change during their lifetime.

The very stable and widespread sex difference in sexual orientation between men and women, even if "rare" exceptions are observed (homosexual subjects), clearly raises the question of the mechanism(s) controlling this difference. It is obvious that, from an evolutionary point of view, this phenotypic characteristic is extremely important since it conditions the reproductive success, sustainability of the species and more directly the transmission to the next generation of combinations of genes that support it (if they exist). Sexual orientation is highly differentiated between males and females in most animal species that reproduce sexually (gender identity does not exist or is impossible to study in animals). One could imagine that the human species has invented a new way, based for example on social learning, to ensure that men and women have an different orientation and therefore a sexual attraction towards the opposite sex, thus ensuring successful reproduction. It is however more parsimonious to consider that the mechanisms existing in animals have been transmitted to humans, even if cultural controls, unidentified so far, have been added.

. . . .

CHAPTER 3

The Origins of Human Sex Differences

3.1   BIOLOGY OR EDUCATION?

If we accept the existence of multiple differences between men and women, this then raises the question of their origin. Do these differences have a biological origin and do they result from the differential expression of genes in men and women? Are they hormonal in nature and induced by the presence during ontogenesis (organizing effects) or in adulthood (activating effects) of different concentrations of estradiol or testosterone? Or alternatively, are they induced by the environment in which the young boy or girl is living? Do education, parents' expectations and social pressures play a significant role in the development of these differences? The answer to this complex issue depends on the difference considered and, as is often the case in biology (and even more in psychology), this response is often mixed, including a part of biological determinism (genes, hormones) and a part of determinism by the environment (education, social environment).

It is clear that the primary morphological sex differences affecting genital structures are due almost exclusively to early hormonal influences. The fusion of genital folds into a scrotum and the development of the genital tubercle in a penis are induced during early embryonic life in animals and in humans by the action on the genital skin of testosterone that must first be transformed into dihydrotestosterone (DHT) by the 5-alpha reductase enzyme to exert these effects.

The determination of other morphological characters is more labile; they are variable during life, unlike external genital structures that are irreversibly determined before birth. While being largely controlled by sex steroids, they are also affected by the environment. One might think here the example of breasts that are a secondary sexual characteristic typical of women, but their growth can be induced in boys by accidental exposure to high quantities of estrogens, for example, following a consumption of poultry meat contaminated with these hormones.

Other differences may be determined more or less equally by biological and environmental factors. For example, the increased capacity of men to metabolize alcohol as compared to women results partly from a differential expression of genes in the liver (in mice 74 % of genes are differently expressed in the liver of males and females) (Yang et al., 2006) and also from the fact that on average men drink more alcohol than women and that enzymes that metabolize the alcohol are inducible.

Finally, gender differences can also reflect more or less exclusively past experiences of the individual. Education, activities and expectations of society and parents indeed vary according to the sex of a child and given the importance of learning in humans, it is not surprising that all these factors have a profound effect on the development of differences between sexes. This is particularly the case for many differences in behavior and cognitive skills, but that does not mean that there is no parallel biological contribution to these differences. We will see later that some aspects of cognitive behavior in men and women can be permanently modified by sex steroid hormones acting during embryonic life. For example, girls that have been exposed prenatally to abnormally high levels of androgens (due to a congenital hyperplasia of adrenal glands) have a masculinized play activity in infancy and are attracted by professional careers that are typically masculine.

3.2   BIOLOGICAL ORIGINS OF HOMOSEXUALITY?

When considering sexual orientation and gender identity, many people favor the idea that these behavioral traits are conditioned by education and early interactions with the parents. Following ideas originally developed by Sigmund Freud (Freud, 1905/1975), it is thus largely accepted in many countries that homosexuality results from an arrest in psychosexual development induced by improper interactions with the father or the mother (see LeVay & Valente, 2006 for a more detailed description). However, evidence for a direct effect of the social or familial environment on sexual orientation and gender identity is at best scarce when it is not negative. Other theories of homosexuality based on early social or experiential factors have also been developed but they also do not seem sufficient by themselves to explain the phenomenon of sexual orientation or gender identify (see LeVay, 2010; Balthazart, 2011 for details).

In contrast, recent research on the neuroendocrine controls of behavior in animals and humans has accumulated over the last few decades, an important amount of information strongly suggesting that prenatal hormones play a key role in determining this sexually differentiated behavioral characteristic. The rest of this presentation will be devoted to a review of these arguments. To fully understand their significance, it is however important to step back for a moment and to summarize what we know concerning the role of sex steroids during early life in the determination of adult behavioral characteristics.

It must be noted that even if, in my opinion, the evidence largely favors a major influence of biological factors in the determination of sexual orientation and identity, this should not be taken as an evidence to negate an effect of currently unidentified experiential factors. Depending on the trait under study, the relative importance of biology (genes and hormones) and of the environment varies enormously, but there is almost always an interaction between these factors. For morphological traits, the part of biological control is obviously more important but it remains that the message transmitted by genes cannot develop in the absence of permissive environmental conditions. For example, the height of an individual is largely determined by genes (parents of tall children are taller than average) but the genetic potential can only be expressed in the presence of an adequate supply of food and good sanitary conditions. This explains the rapid growth of the average stature of the population in Western Europe during the last century. The average size of males was 1.66 m in 1900 and has reached 1.75 m at the end of the 20th century. This development has greatly accelerated between 1960 and 1990, and 5 cm in average size has been gained in 30 years. Such a difference cannot, of course, be the result of genetic evolution in such a short period of time (100 years or about 5 generations). It is the result of improved living conditions and especially the living conditions of young children (health care, diet, . . .).

·   ·   ·   ·

CHAPTER 4

How Can Effects of Biological Factors on Sexual Orientation Be Studied in Humans?

For various fairly obvious technical and ethical reasons, the experimental analysis of the neurobiological and endocrine bases of sexuality is difficult in humans and must often be limited to the study of clinical cases. Direct manipulations of brain physiology are prohibited for ethical reasons. Similarly, experiments involving removal of the source of gonadal steroids (castration or ovariectomy) and their replacement by exogenous steroid regimes cannot be performed for experimental purposes and are only undertaken when there is a medical indication that will complicate the analysis of the results. Finally, interviewing people on their sexuality is fraught with difficulties not to mention actually observing direct sexual interactions.

Most of the data that can be collected in this field are therefore essentially correlative, and their interpretation remains usually problematic. It is often impossible to ascertain whether a difference in behavior that is shown to be associated with a clinical endocrine or neural condition is causally induced by this condition or is simply a by-product that has no direct causal link. The confirmation of a specific connection can only come indirectly from the comparison with true causal studies performed on animals. Three complementary strategies are thus commonly employed in studies designed to understand biological controls of human sexual orientation.

The first one involves the careful analysis of the so-called experiments of nature that are in fact clinical cases in which either a genetic or an endocrine perturbation has been identified and medical doctors assisted by psychologists and psychotherapists try to relate the biological condition to its behavioral consequences. The obvious drawback of this approach is that these experiments are of course only pseudo-experiments so that is it often difficult if not impossible to identify with certainty the specific character that is the cause of the observed phenotype.

A second alternative strategy that has been used widely is based on the comparative analysis in homosexual and heterosexual populations of various morphological, physiological or behavioral traits that are known to be sexually differentiated in control heterosexual populations. These studies focus especially on traits that are known or at least suspected to differentiate based on prenatal actions of sex steroids. The reasoning here is that if these traits are different in homo- and heterosexual populations, this might reflect a differential exposure to sex steroids in these populations.

It is clear however that these two approaches, even in their most sophisticated application, only provide at best correlative data that remain difficult to interpret in the absence of true causal experiments. The third strategy therefore requires causal manipulations, but these manipulations are usually impossible in humans for pragmatic or ethical reasons that were already mentioned before. They have therefore to be performed in animal models that will be able to provide the background causal data needed to interpret the correlative human results. To allow the reader to develop a broader understanding of the human data supporting the role of biological, in particular endocrine, factors in the genesis of homosexuality, we shall thus provide here a brief overview of what has been clearly established in various animal models concerning the role of hormones in the control of behavioral sex differences.

. . . .

CHAPTER 5

Endocrine Control of Sex Differences

5.1   ACTIVATION OF SEXUAL BEHAVIORS BY SEX STEROIDS

Like in humans, many behaviors in animals are sexually differentiated and are produced preferentially or exclusively by one sex. For example, when a male and a sexually receptive female rat are placed in a same enclosure, they will interact sexually but males will show a series of stereotyped behaviors including mounts, intromissions and pelvic thrusts eventually culminating in ejaculation, whereas females will display a number of proceptive behaviors such as ear wiggling or hopping and darting before they adopt the lordosis posture that will allow males to intromit. Such differences in the behaviors displayed by males and females in the context of reproduction (and also in other contexts) are seen in most if not all animal species. In the early days of behavioral endocrinology (before the middle of the twentieth century), it was assumed that these behavioral differences resulted from the presence of different hormones in males and females, specifically testosterone in males and estradiol/progesterone in females. These hormones are indeed required to allow expression of these behaviors. If animals are gonadectomized (castration in males and ovariectomy in females), the behavior will no longer be observed but it will be restored by a treatment with these hormones. It is said that the steroid hormones activate the behaviors. They do so largely by affecting the transcription of specific genes and the resulting proteins modulate neurotransmission in specific neural circuits controlling the behaviors (Becker et al., 2002; Pfaff et al., 2002; Etgen & Pfaff, 2009; Nelson, 2011). These effects usually require a few days of exposure to the hormone to develop, but they are essentially transient. When exposure to the hormone is discontinued, behavior expression regresses and eventually disappears.

As pure steroid hormones (essentially testosterone and estradiol) became available for experimental purposes, it rapidly became clear that the specificity of male and female behavior was not only the result of the exposure to different hormones. Indeed, testosterone often failed to activate male-typical behaviors in females and similarly the sequential treatment with estradiol and progesterone that activates lordosis in females did not induce expression of this behavior in males. Furthermore, some male-typical behaviors were also activated by estradiol in males. This observation eventually lead to the formulation of the aromatization hypothesis according to which the activation of male copulatory behavior depends on the action at the cellular level of estradiol (the typical "female" hormone) produced by aromatization of testosterone in the brain (Beyer et al., 1970; Naftolin et al., 1971; Naftolin et al., 1975). This hypothesis was later demonstrated to be true in many species. Thus, it is not the type of hormone that determines the type of behavior that is produced but rather it is the nature of the neural substrates on which this hormone acts (sex of subject).

5.2   SEXUAL DIFFERENTIATION (ORGANIZATION) OF BEHAVIOR

It is now firmly established that the differentiated behavioral responses to steroids of males and females are often the result of early actions of these same steroids. During ontogeny, the brain, which was initially neutral, differentiates into a male or female brain. This differentiation occurs very early in life during the embryonic period or just after birth, and it is completely irreversible. In mammals, early exposure to testosterone produces a male phenotype: the behavioral characteristics of the male are strengthened (masculinization) and the ability of males to show behavior typical of females is reduced or lost (defeminization). The female phenotype apparently develops in the absence of hormone (or in the presence of very low levels of estrogen) (McCarthy & Ball, 2008). These processes that take place in physiological conditions during ontogeny have been experimentally reproduced in a variety of animal species by manipulations blocking steroid action or by injecting steroids in embryos or neonates, as illustrated for the rat in Figure 4.

As illustrated, the injection of testosterone in an embryonic female rat produces a profound masculinization of her behavior. This genetic female is capable as an adult to respond to injections of testosterone by producing the full range of sexual behaviors typical of males. Conversely, if a newborn male is castrated at birth, his behavior will not be fully masculinized and he will be unable in adulthood to achieve with high-frequency behaviors typical of his sex. Correlatively, his sexual behavior is not defeminized: he is capable of producing the posture of sexual receptivity (lordosis) in response to a sequential treatment with estradiol and progesterone. Thus, the behavioral phenotype of rats can be completely reversed by early hormonal manipulation. One can produce individuals who will, in adulthood, exhibit the behavioral characteristics of males or females regardless of their genetic sex (Goy & McEwen, 1980; McCarthy & Ball, 2008).

It is also important to note that as shown in Figure 4, the masculinization/defeminization of sexual behavior may be obtained in female rats by prenatal injection of estradiol instead of testosterone. This observation stems from the fact that testosterone masculinizes behavior during ontogeny largely through its conversion into estradiol (by aromatization in the brain). This notion is confirmed by the fact that if a genetic male, whose embryonic testis produces large quantities of testosterone is injected during the last week of uterine life and the first week of postnatal life with an aromatase inhibitor, i.e., a compound that pharmacologically blocks the conversion of testosterone to estradiol, sexual behavior will not be fully masculinized. It therefore appears that the effects of testosterone in

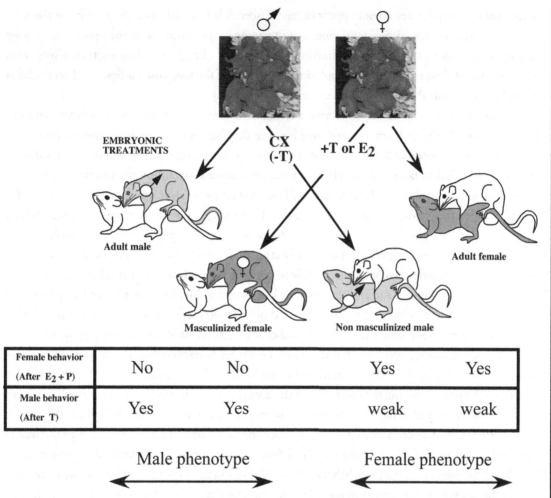

Female behavior (After E_2 + P)	No	No	Yes	Yes
Male behavior (After T)	Yes	Yes	weak	weak

Male phenotype   Female phenotype

FIGURE 4: Schematic representation of the hormonal processes during ontogenesis that induce sexual differentiation of brain and behavior. CX: castration, T: testosterone, E2: estradiol, P: progesterone.

rats are produced at the cellular level by estradiol derived from aromatization, both during ontogeny (masculinization of behavior) and during adulthood (activation of sexual behavior).

Given that the ovaries of pregnant females produce high amounts of estradiol that passes the placental barrier easily, one might wonder why maternal estrogens do not also masculinize female embryos. However, the plasma of rat embryos contains a protein called alpha-fetoprotein (AFP) that is secreted by the liver and binds with high affinity to estrogens thus preventing them from entering the brain (Bakker et al., 2006). In contrast, AFP does not bind to testosterone, which can thus reach the male brain where it exerts its effects after being locally aromatized into estradiol.

Some species like primates do not appear to have discernible levels of circulating AFP but the relative importance for sexual differentiation of testosterone aromatization in the brain is also quite variable from species to species. In monkeys, for example, effects of androgens themselves seem more important than estrogens. Extrapolation to humans of the mechanisms described on the basis of studies in animals should thus be made with due caution.

Two other aspects of the endocrine mechanisms that control the sexual differentiation of behavior must absolutely be mentioned here because they have a very important impact on the interpretation of homosexual behavior and its possible hormonal origin which will be discussed later. It is indeed critical to note that the effects of testosterone or estradiol on the behavioral sex of an individual only occur during a limited and well-defined period of development of the animal. This period is called the "critical period" and corresponds to a stage of brain development during which the brain is still very plastic so that it can be profoundly changed by exposure to sex steroids. In rats, a species in which the young are born at an early stage of development (they are naked, blind, do not regulate their body temperature and are unable to move by themselves), this period essentially covers the last week of embryonic life and the first week of postnatal life. In other species like pigs where the young are more developed at birth, the period of brain development takes place entirely during embryonic life and, therefore, the critical period of sexual differentiation is entirely prenatal.

These organizing effects of steroids are also completely **irreversible**. If testosterone or estradiol are present during the critical period and masculinize and/or defeminize sexual behavior, these effects will be present during the entire life of the animal. A female who has been masculinized and defeminized by early exposure to testosterone (or estradiol) is unable during her whole life to present the behavior of lordosis in response to an appropriate treatment by estradiol and progesterone in adulthood. Correlatively, this masculinized female responds to treatment with testosterone in adulthood by showing copulatory behavior typical of the male. This reaction, normally characteristic of the male, remains present in these females masculinized throughout their life. Conversely, a male castrated immediately after birth retains during his entire life the possibility to show lordosis behavior, but will never fully respond in adulthood to a treatment with testosterone that is supposed to activate male-typical behaviors. Contrary to the activating effects of steroids that are fully reversible and are only present during the period of exposure to hormones, the organizing effects are permanent and last for the entire life of the animal.

Similar mechanisms of hormonal control of the sexual differentiation of behavior have now been identified in many species of mammals and birds. They appear to be quite widespread at least in homeothermic vertebrates even if there are fine differences in the mechanisms involved. For example, the relative importance of the processes of masculinization and defeminization varies from one species to another. Sexual differentiation of male rats involves an almost complete defeminization linked to a limited behavioral masculinization, whereas in rhesus monkeys there is rather a marked masculinization of behavior that is not associated with the important process of defeminization.

In summary, the female phenotype develops in mammals spontaneously in the (relative) absence of hormones but testosterone (eventually through aromatization) is required to impose masculinity. This differentiation process takes place spontaneously during early development and is completely irreversible. Importantly, these principles identified in animal studies appear to be applicable mutatis mutandis to humans. Differences between species lie in the details.

5.3   SEXUALLY DIFFERENTIATED STRUCTURES IN THE BRAIN

When the process of sexual differentiation of reproductive behavior was first identified in the late 1950s (Phoenix et al., 1959), it was generally accepted that the brains of males and females are morphologically identical. Consequently, the sexual differentiation of behavior was thought to be mediated by small irreversible changes in brain physiology (e.g., in the function of specific neurotransmitter systems) in the absence of any significant modification of brain morphology. This view changed markedly during the next two decades following the discovery, first of minute subcellular sex differences in the mammalian brain (Dörner & Staudt, 1969; Raisman & Field, 1971) and then subsequently of larger differences in the volume of several brain nuclei specifically implicated in the learning and production of singing behavior in birds of the group called passerines or songbirds (e.g., canaries and zebra finches) (Nottebohm & Arnold, 1976). These nuclei (dense clusters of neurons) have a volume that is 2 to 5 times larger in males than in females so that the sex difference can be seen with the naked eye in stained brain sections.

This discovery of large neuroanatomical sex differences in songbirds prompted a reanalysis of this question in mammals. Roger Gorski and his group at the University of California in Los Angeles soon discovered the existence of a "sexually dimorphic nucleus" (SDN) in the preoptic area of rats (Gorski et al., 1978). Like in canaries, this volumetric difference is also very important: the SDN of the male rat is five times larger than the SDN of females (Figure 5).

The location of this nucleus in the middle of the preoptic area, which is the area of the brain involved in a privileged way in the activation by testosterone of male sexual behavior (Balthazart & Ball, 2007) strongly suggests that this nucleus plays a key role in controlling this behavior. Many experiments have been conducted to test this idea but have obtained only a partially positive answer. It was shown that lesions in the preoptic area covering all or part of the SDN inhibit the expression of male copulatory behavior but that, in contrast, more discrete lesions localized specifically at the SDN were usually without effect (Arendash & Gorski, 1983). A few studies, however, contradict this conclusion and show that SDN lesions inhibit male-typical sexual behavior activated by testosterone in males (De Jonge et al., 1989) and females (Turkenburg et al., 1988). According to the authors, the effectiveness of lesions in their studies would relate to the fact that they used sexually

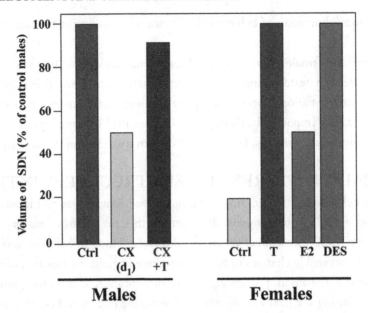

FIGURE 5: The volume of the sexually dimorphic nucleus (SDN) of the preoptic area of rats is about five times larger in males than in females and this difference is induced by steroids during the perinatal life. Castration (CX) in the first day after birth (d1) reduces nucleus size in males and this may be offset by treatment with testosterone (T). Perinatal treatment of females with testosterone, estradiol (E2) or a synthetic estrogen (DES) increases the volume of this nucleus as will be later observed in adulthood. According to Gorski (1984).

inexperienced subjects. The SDN and surrounding preoptic area are thus involved in the expression of male sexual behavior but the specific lesion of SDN is generally insufficient to inhibit behavior.

The analysis of the hormonal mechanisms that control the volume of the rat SDN also provides additional arguments, suggesting a causal relationship between the size of this nucleus and behavioral activity. The size of the male rat SDN does not depend on the hormonal status of animals in adulthood but, it is, like the behavior, determined by the endocrine environment experienced by the animal during the perinatal life. The injection of testosterone in young females during this critical period produces adult females displaying a size of the SDN identical to that of a male (which is five times greater than in untreated female). Conversely, if a male pup is castrated at birth, the size of his SDN in adulthood is significantly lower than that of a typical male (about half) (Jacobson et al., 1981; Arnold & Gorski, 1984). His SDN is however larger than in a female because the differentiation of the SDN is induced by testosterone in the last week of prenatal life and the first postnatal week. Castration at birth thus interrupts a process that is already underway and the size of the SDN is already greater than in a female who has never been exposed to testosterone (see Figure 5). Once the volume of this nucleus is determined at the end of the first week of postnatal life, it is fixed for the rest of life: like the type of sexual behavior (male or female) that can be displayed, this feature can no longer be changed by hormone treatments in adulthood.

It should be noted that the "masculinization" of the SDN volume by perinatal actions of testosterone is induced at the cellular level by estrogens produced in the brain by aromatization of testosterone. This conclusion is amply demonstrated by the fact that injections of estrogens to a young female rat increase in the same way as testosterone the SDN volume (if the injected dose is sufficient to saturate the binding capacity of alpha-fetoprotein, or if a specific estrogen such as diethylstilboestrol, DES, which does not bind to alpha-fetoprotein is used). Furthermore, injection of an aromatase inhibitor blocks the effects of testosterone on the masculinization of the SDN volume. This control by aromatization of the SDN volume thus represents another correlation with sexual behavior, which also differentiates under the action of estrogens produced in the brain by aromatization.

Following this discovery of the SDN in rats by Roger Gorski and his colleagues, similar sexually dimorphic structures were discovered in the preoptic area of many other species of mammals including humans (see Chapter 11), but also birds, reptiles and amphibians (Tobet & Fox, 1992; Balthazart & Ball, 2007). These dimorphic structures of the preoptic area are not always homologous in different species and are not necessarily controlled by the same hormonal mechanisms. For example, the larger size in males of the preoptic SDN in some species such as the Japanese quail is the result induced by the higher concentration of testosterone in adult males as compared to females (Panzica et al., 1987). In technical terms, this difference is linked to a differential activation by steroids in adulthood rather than to a differential organization during early life. It seems, however, that in most mammals studied to date the sex difference in SDN volume results from organizational effects of sex hormones exerted during early life, either pre- or immediately postnatally.

Finally, it should also be mentioned that the identification of an SDN in the preoptic area in many species initiated an active line of research that has led during the following decades to the identification of numerous morphological sex differences in other parts of the brain. A large number of brain structures are indeed either larger in males than in females (e.g., the nucleus of the stria terminalis in rats) or larger in females than in males (e.g., the ventro-medial nucleus of the hypothalamus). It is beyond the scope of this presentation to consider all these neuroanatomical sex differences in animals but we will return to this topic in more detail in the discussion of the morphological differences between brains of men and women (Chapter 11).

•   •   •   •

CHAPTER 6

Organizational Effects of Steroids on Sex Partner Preference in Animals

Most animal studies devoted to sex differences affecting sexual behavior have focused until quite recently on the differences in the type of behavior patterns expressed by the two sexes. These studies are therefore of little interest for the understanding of human behavior in which patterns of behavior expressed by men and women during sexual interactions are little or not different (see Chapter 1). Sexual relations between men and women occur in a variety of positions that reflect cultural influences but are neither stereotyped nor probably controlled by specific biological or hormonal mechanisms. The main differences between men and women in the field of reproductive behavior indeed concern sexual orientation and sex (gender) identity or role that are the focus of this presentation.

Sexual identity and sex role are specific to men and women and cannot be studied in animal models (it is not possible to ask an animal what is his or her sex). In contrast, sexual orientation can be easily studied in animals by offering them a choice between a male or female sexual partner. The observer can then record toward which of these partners the test animal will orient his/her sexual behavior. This type of research is not as developed as the research on sexual behavior *sensu stricto* and it was started more recently. It has however provided important insights into the origins of human homosexuality. It is indeed firmly established at this stage, in rodents at least, that the sexual orientation of reproductive behavior is controlled by the action during ontogeny of the same hormones that control the activation of sexual behavior and that these hormones act in the same brain regions to activate sexual behavior and determine its orientation. We summarize these data in the following pages as they establish the theoretical context from which we propose thereafter an explanatory model of human homosexuality based on early (embryonic) and irreversible effects of sex steroids.

6.1 PARTNER PREFERENCE IN MALES IS CONTROLLED BY THE PREOPTIC AREA

In animals as in humans, most sexual activities of the male are oriented towards females and vice versa. This orientation can be quantified in standardized laboratory conditions with the help of three compartments cages (see Figure 6).

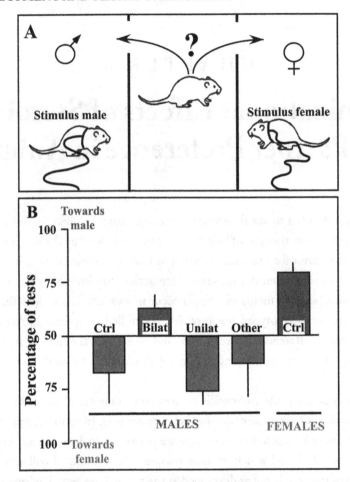

FIGURE 6: Experimental three compartments cages used to measure sexual preferences in small rodents and in ferrets (A) and effect of lesions of the medial preoptic area on these preferences in the ferret (B). Male control (Ctrl) ferrets spend the majority of their time in the compartment containing a female and vice versa. Male preferences are reversed following a bilateral lesion (Bilat) of the preoptic area but not by unilateral lesions (unilateral) or lesions of a different site of the brain (Other). Panel B drawn from the data in Paredes & Baum (1995).

The test animal is placed in the central compartment of the cage and male and female stimuli are randomly placed in the left and right compartments. They are maintained by a harness that prevents their passage into the central cage, while allowing the test animal to freely visit the three compartments (Figure 6A). This experimental procedure was used for a variety of animal species but the most detailed data are available for the laboratory rat (*Rattus norvegicus*) and the ferret

(*Mustela putorius*). It was observed, as might be expected, that in such a device males spend most of their time (often more than 80–90%) in the compartment containing a sexually receptive female, with whom the male will mate. He will visit for much less time the compartment containing a male and in his presence he will exhibit only investigations or aggressive behaviors. In contrast, a sexually mature and receptive female will spend most of her time in the compartment containing a male with whom she will mate even if he is held captive in his compartment.

The preference for the compartment containing a subject of the opposite sex is activated by steroid hormones, which, in parallel, modulate the expression of sexual behavior. Castration of males or female ovariectomy eliminates or greatly reduces this preference.

A limited number of studies have also investigated the neural sites underlying the expression of these preferences in males and show that male sexual preferences are controlled by the medial preoptic area. The first experiment supporting this conclusion was performed in ferrets by Michael Baum and his colleagues at Boston University (Figure 6B). Like rats, ferrets have a sexually dimorphic nucleus in their medial preoptic area. The volume of this nucleus is significantly larger in males than in females. After bilateral electrolytic lesions of this nucleus, males who previously were spending most of their time in the compartment containing a female display a nearly complete reversal of this choice and spend during the post-lesion tests most of their time in the chamber containing a stimulus male (Paredes & Baum, 1995). They therefore show a partner preference that might be described as homosexual or at least bisexual in humans.

These lesion effects are anatomically specific since lesions of the same size that entirely missed their target or missed it on one side of the brain do not affect the partner preference of the males: they continue to show a preference for the compartment containing a sexually receptive female. Only bilateral lesions placed exactly at the level of the sexually dimorphic nucleus of the preoptic area are able to reverse the sexual orientation of males. Similar results have been observed in rats (Paredes et al., 1998).

6.2   EARLY STEROID EXPOSURE DETERMINES PARTNER PREFERENCE

The three-compartment cage was also used to investigate how sexual partner preferences develop during ontogeny. We have previously seen that embryonic testosterone is responsible for the masculinization/defeminization of reproductive behavior in rats (Chapter 5). Injection of an aromatase inhibitor during the end of embryonic life or during the first week of postnatal life similarly inhibits masculinization of behavior thus indicating that testosterone is acting largely through its estrogenic metabolites. Quite interestingly, the same treatment with an aromatase inhibitor also blocks the development of the heterosexual partner preference in male rats (Bakker et al., 1993; Bakker et al., 1996b). When adults, rats that were treated in early life with an aromatase inhibitor spend, unlike normal rats, the majority of their time in the compartment containing another male and largely ignore the compartment containing a sexually receptive female (Figure 7A–B). Moreover, these rats rarely attempt to mate with sexually receptive females that are present and even allow stimulus males to mount them. They are thus showing a form of partner preference that would be qualified, in humans, as homosexuality or at least bisexuality. Later experiments indicated that inhibition of aromatase during the first three postnatal weeks is more effective in reversing partner preference than prenatal inhibition (Houtsmuller et al., 1994) (Figure 7C–D).

In addition, these rats have, as adults, a small SDN in the preoptic area, that is characteristic of the female (Houtsmuller et al., 1994). It therefore appears that the absence of estrogen exposure of male rats during the perinatal life has a lasting effect on the sexual preferences of the male in adulthood. These rats are not masculinized and tend to prefer other males to females.

Quite interestingly, the brain of these "homosexual" male rats responds in an atypical manner to the presentation of olfactory stimuli of a sexual nature. In control animals, the nuclei that process olfactory information related to sexual behavior are activated in males by presenting female-typical odors (e.g., cage bedding soiled by the urine of females) while the female brain is activated by the odor of males. In contrast, the brain of a male is not activated by the presentation of a litter soiled by a male and vice versa. This is very different in male rats treated during the perinatal period with an aromatase inhibitor: their brain is highly activated by the odor of other males (Bakker et al., 1996a). The experimental manipulation of the early hormonal environment has produced a profound change in what the animal sees as sexually attractive or exciting.

Recent studies also showed that the treatment (during the first three weeks of life) of young female rats with an estrogen (estradiol benzoate) has the opposite effect. This treatment increases their preference for females, a preference that would be classified as homosexual in women (Figure 8) (Henley et al., 2009). These preference tests were conducted at the adult age in ovariectomized females exposed to various hormone treatments (estrogen alone or estrogen plus progesterone) to activate different aspects of sexual behavior. The preference reversal was observed in both hormonal conditions, which suggests that "homosexual" preferences induced by perinatal treatment cannot be modified by hormones in adulthood.

These observations initially performed in rats were confirmed in mice although in this species androgens themselves seem to play a more important role in the differentiation of sexual partner preferences. Sexual differentiation is deeply disturbed in mice suffering from a mutation (called tfm) of the androgen receptor that prevents the action of testosterone or DHT (Bodo & Rissman, 2007). In adulthood, these tfm male mice, when gonadectomized and treated with estrogens to activate sexual behavior and motivation, have responses similar to those of females. For example, they spend more time investigating like females bedding soiled by the urine of males, whereas control males prefer bedding soiled by the urine of females. More females and tfm males showed no preference

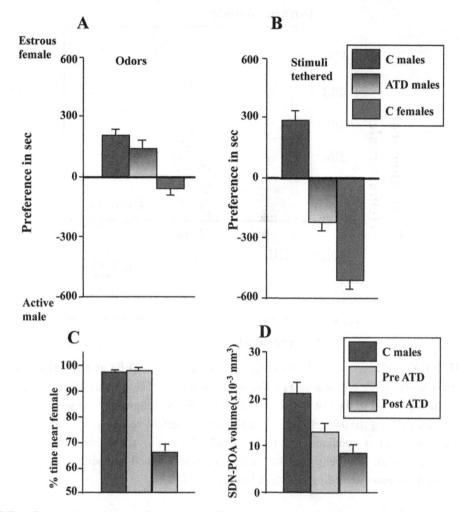

FIGURE 7: Perinatal inhibition of aromatase affects the adult sex partner preference of male rats and correlatively decreases the volume of the sexually dimorphic nucleus of the preoptic area (SDN-POA). **A–B.** Relative partner preference expressed as time spent in the female compartment minus time in the male compartment when subjects were tested with sexual odors only (A) or with tethered live subjects as stimuli. Experimental subjects were control (C) males and females and males exposed to the aromatase inhibitor ATD pre- and post-natally (Bakker et al., 1996b). **C–D.** Percentage of time spent in the female compartment and volume of the SDN-POA in control (C) males and in males that had been exposed pre- or postnatally to the ATD (Houtsmuller et al., 1994). Redrawn from data in (Bakker et al., 1996b) and (Houtsmuller et al., 1994).

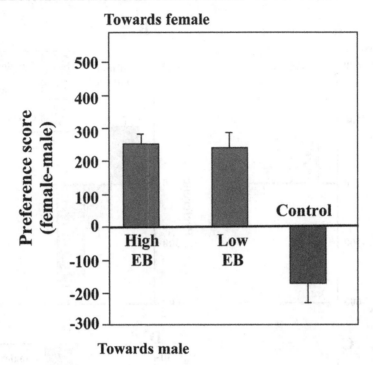

FIGURE 8: Effect of treatment with an estrogen, estradiol benzoate (EB), at high or low dose, during the first three weeks of life on the sexual preferences of female rats. The preference score represents the time spent by the animal in the test chamber containing a female minus the time spent in the chamber containing a male during the first test performed in adult ovariectomized subjects treated with estradiol. A negative score indicates a preference for males (usually observed in control females). A positive score indicates a reversal of this choice (preference for females) was observed following treatment with EB at both doses used. Redrawn according to data in Henley et al. (2009).

for a partner of one sex or another in simultaneous choice tests whereas control males showed a strong preference for a female partner. Finally, exposure to bedding soiled by males (but not to clean bedding) induced neuronal activation in the preoptic area and the nucleus of the stria terminalis (measured by an increased expression of the *c-fos* gene) in females and tfm males but not in control males. The same researchers also demonstrated that treatment of neonatal females with an androgen (dihydrotestosterone) masculinize for life all these behavioral features and the response of the nervous system to male odors (Bodo & Rissman, 2008). Estrogens are however also implicated in mice as they are in rats (Bakker et al., 2002b; Brock & Bakker, 2011). These findings thus support the idea that sexual partner preferences are controlled during ontogeny by the action of sex steroids, testosterone and its estrogenic metabolite estradiol.

Importantly, it cannot be over-emphasized that these early hormonal manipulations have, as far as we know, completely irreversible effects on both the type of sexual behavior that will be displayed in adulthood (male- or female-typical) and its orientation ("homo- vs. heterosexual"). In particular, these behavioral characteristics cannot be modified by hormonal treatments performed during adulthood. Principles derived from these animal studies play a crucial role in the analysis of the potential endocrine controls of human sexual orientation.

6.3   HOMOSEXUAL SHEEP

Studies described so far concern animal models in which some aspects of what would be called a homosexual orientation in humans have been experimentally induced by brain lesions or early hormonal manipulations. In all cases, however, these subjects almost never displayed an exclusively homosexual orientation, they were in most cases bisexual with a more or less strong homosexual preference. This homosexual orientation was also experimentally induced as opposed to human homosexual orientation that develops spontaneously. Over the past 10 years, a spontaneous model of exclusive homosexuality has, however, been described in sheep living in the North Western part of the USA and deserves to be considered here in detail.

It was noted during standardized behavioral tests that a significant fraction of male sheep in this population (8%) shows little or no reaction in the presence of females but are nevertheless not asexual: they display active sexual behavior toward other rams. Furthermore, these males mate exclusively with other males when given a simultaneous choice between a male or female partner. The sexual motivation and sexual performance of these males is thus not in question. It is specifically the orientation of their behavior that is atypical.

Various factors that could explain this male-directed sexual behavior such as the same-sex rearing which is common in sheep or genetic determinants were first ruled out (see Roselli et al., 2011 for review) and studies then focused on endocrine and neural aspects of this preference for males. Anatomical studies identified in the sheep preoptic area a sexually dimorphic nucleus similar to the dimorphic nucleus described in rats (Roselli et al., 2004a; Roselli et al., 2004b). This ovine sexually dimorphic nucleus (oSDN) is approximately three times larger in males than in females and contains about four times more neurons. Analysis of this nucleus in male-oriented rams (MOR) showed that it was significantly smaller than in female-orientated rams (FOR), that it contained fewer neurons and it expressed aromatase at a reduced level. The volume of oSDN can also be measured by quantifying, in successive sections, the surface of dense expression of aromatase mRNA. This confirms measurements of the oSDN obtained by conventional histological stains: the oSDN is larger in FOR than in females and MOR have an oSDN smaller than the FOR.

All these characteristics make the oSDN of MOR quite similar to the oSDN of females and differentiate it from the oSDN of FOR. The features of this nucleus therefore correlate with sexual

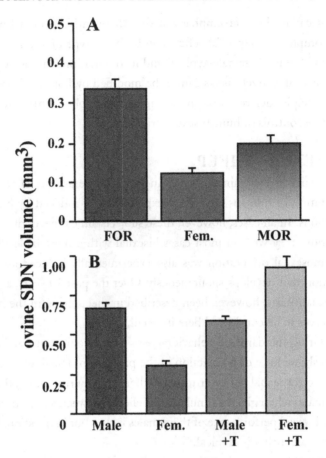

FIGURE 9: Volume of the sexually dimorphic nucleus of the preoptic area (oSDN) in sheep. A. The oSDN volume is larger in female-orientated rams (FOR) than in females (Fem.), but the male-oriented rams (MOR) have a volume similar to that of females. B. Effect of treatment with testosterone between days 30 and 90 of gestation on the volume of oSDN measured at 135 days of gestation (birth at approximately 150 days). A significant increase in the volume of oSDN is observed among females but the same treatment has no effect in males. Redrawn from data in Roselli et al. (2004b) and Roselli et al. (2007).

orientation: subjects attracted to males (females and MOR) are similar and distinguished from subjects attracted to females (FOR) (Figure 9) (Perkins & Roselli, 2007) (Roselli et al., 2011).

The small volume and the other feminized characteristics of the oSDN in MOR could technically be a consequence of their orientation since the adult brain is still plastic and responds to the social environment. However, analysis of the endocrine controls of the volume of this nucleus rather suggests that this volume is determined before birth and thus before sexual behavior expression is initiated. If rams and ewes are gonadectomized in adulthood (castration of males and ovariectomy of females) and after a month without hormone, all subjects are treated for three weeks with testosterone, at autopsy the size of their oSDN is not affected by comparison with control subjects. The oSDN of FOR remains about twice as large as in females and MOR have an oSDN significantly smaller than FOR. These hormonal manipulations performed in adulthood thus do not affect the oSDN volumes, which suggests that this volume is determined during ontogeny, an idea that has received experimental support (Roselli et al., 2007). During the embryonic life of the sheep, which lasts about 150 days, there is between the 50th and 100th days a peak in the concentration of circulating testosterone in males that is not present in female embryos. This peak of testosterone is likely to differentiate the brain and the behavior. Histological analysis showed that in late gestation (days 135 to 140), the oSDN is already clearly discernible and its volume is significantly greater (approximately 60%) in males than in females. This difference is presumably caused by the early peak of testosterone.

Roselli and colleagues accordingly demonstrated that if females are treated with exogenous testosterone between 30 and 90 days of gestation, they display at the end of embryonic life a masculinized oSDN (Roselli et al., 2007). Therefore, the small size of oSDN in MOR is presumably determined before birth and before subjects had an opportunity to express their sexual orientation. Given that this nucleus is located in the center of the preoptic area, a region involved in the control of male sexual orientation, there is every reason to believe that the little oSDN of MOR is (one of) the cause (s) of their atypical sexual orientation resulting from an inadequate masculinization by testosterone during embryonic life.

All available data thus indicate that the volume of the sheep SDN is determined, as in rats, by embryonic exposure to testosterone and cannot be changed in adulthood even by major hormonal manipulations (gonadectomy combined treatment by testosterone). This invariance in adulthood and this early determination indicate that the size of the oSDN is determined well before the beginning of the expression of sexual behavior and orientation. It is thus impossible to imagine that the small size of the oSDN in MOR is the consequence of their behavior. Either it is the cause of the sexual orientation, or alternatively it is the signature of an embryonic endocrine event that would have determined independently the homosexuality of the males and the size of their oSDN.

In conclusion, animal studies demonstrate that sexual partner preference in animals is a sexually differentiated feature like other sexually differentiated behavioral or morphological characteristics. Male-typical partner preference is controlled at least in part by the preoptic area (like sexual behavior; data are not available for females) and partner preference in both sexes differentiates under the influence of pre/peri-natal sex steroids.

·   ·   ·   ·

CHAPTER 7

An Endocrine Model of Human Homosexuality

Recent research has thus identified neuroendocrine processes that control the differentiation and activation of many aspects of behavior in animals. Many questions remain open, particularly how the hormones act at the cellular level in the brain, but this animal work nevertheless identified a critical principle that governs the development of sex differences in behavior and in particular in the sexual partner choice.

This aspect of sexual behavior is largely controlled by sex steroids acting during the end of embryonic or early postnatal life to determine, in an apparently irreversible manner, the sex of subjects that will be considered as suitable sexual partners during the rest of the animal's life. Most of this work relates to the male's partner preference but more recently Henley and collaborators demonstrated an increase in preference for females congeners in female rats injected with estradiol benzoate during the first three weeks of postnatal life (Henley et al., 2009). This suggests that similar principles may apply to females, even if more experimental work is needed on this topic.

This experimental animal work suggests a theoretical model of control of human sexual orientation similarly based on an early action of sex steroids (mostly testosterone but also possibly its estrogenic metabolites) that would determine or at least induce a major predisposition for a given sexual orientation to be displayed later in adult life (see Figure 10).

The male-typical attraction for women would develop, according to this model, following exposure to a critical concentration of testosterone during a, yet to determine, critical period of early development. By default, in the absence of this exposure to testosterone, a female-typical attraction for men would develop. Because in a natural population, all biological characteristics vary greatly between individuals (as illustrated in Figure 10 by the inverted U-shape curves of distribution), one could then expect that some male individuals at the lower end of the distribution curve for testosterone concentration would actually experience testosterone concentrations that are below the critical threshold (vertical line in the middle of the figure). As a consequence, they would develop a female-typical attraction for men and thus develop into homosexuals. Conversely, if female embryos

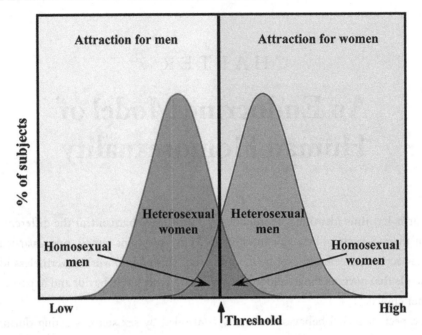

FIGURE 10: Theoretical model illustrating how "random" fluctuations around the average concentration (or action) of testosterone present in male and female embryos could affect sexual orientation. Male embryos at the low end of the testosterone distribution and female embryos at the high end of the distribution are in this model exposed to testosterone concentrations typical of the opposite sex and develop a homosexual orientation.

are exposed to a concentration of testosterone above the critical threshold, they will develop a male-typical attraction for women and later become lesbians. Recall indeed that a homosexual attraction is in fact an attraction that is similar of the attraction usually displayed by the other sex even if the specific individuals that will be found attractive (by homosexual men and women on the one hand and by homosexual women and men on another hand) will probably be quite different. Homosexual subjects would in this model simply be the individuals experiencing endocrine conditions during early life that are at the low end (homosexual men) or at the high end (homosexual women) of the distribution of testosterone action.

I am using the word action here quite intentionally because it is obviously not only the circulating concentration of testosterone that matters. To exert its effects on morphology, physiology and behavior, testosterone must indeed reach its targets (mostly the brain when behavior is concerned) and bind to specialized proteins called receptors that will act at the genomic level to modulate

DNA transcription and thus cell function. If these receptors are lacking or (partly) non-functional as happens in some clinical conditions such as the androgen insensitivity syndrome (AIS) caused by mutations of the corresponding genes, testosterone action will be impossible or will be variably suppressed. In the model, failure to develop a male-typical sexual attraction for women could thus also take place in the presence of normal concentration of circulating testosterone if the receptor for this hormone is partly or completely non-functional.

It should also be noted that even if the model illustrated in the figure highlights testosterone action as the critical factor influencing sexual orientation, research in rodents indicates that testosterone exerts a large part of its action after being converted into an estrogen by the enzyme aromatase. There is evidence suggesting that in primates the role of estrogens is not as important as in rodents when compared to the role of androgens *per se* in the control of sexual differentiation. One might thus hypothesize that the same is true in humans and that androgens, not estrogens, play the major role in sexual differentiation but at this time no firm conclusion can be drawn on this question.

Even with these limitations, this model provides a specific theoretical framework that can be tested against available evidence derived from human research. The key question then becomes: is there any evidence from clinical and epidemiological studies indicating that endocrine mechanisms controlling the differentiation of sexual partner preference in animals are still active in humans and contribute to explain the homo- or heterosexual orientation of adult subjects? One could argue that this is presumably the case given the evolutionary continuity observed in mammals. In addition, given the extreme importance of sexual orientation from an adaptive point of view (homosexual subjects do not reproduce or at least have a lower number of children), it could be expected that exceptionally high selective pressures must be associated with this behavioral trait. Consequently, established mechanisms that have proved to be effective in our ancestral species (probably all birds and mammals) should have been conserved instead of evolving a set of new, probably more volatile (and thus variable), mechanisms for the control of sexual orientation that would be based on education as suggested by Freudian theories, behaviorism or social constructivism.

· · · ·

CHAPTER 8

Sex Steroids Modulate Sexual Motivation during Adulthood in Humans . . . But Not Sexual Orientation

It is important in this context to first recall that sex steroids (testosterone, estradiol, progesterone, . . .) that are observed in all vertebrates are also circulating in humans and do so in concentrations that are very similar to the concentrations present in other mammals. The brain regions that control sexual behavior in rodents and primates have also changed very little between these mammals and humans. The same nuclei (clusters of neurons) can be found in the preoptic area, the hypothalamus and limbic system. These nuclei in the human brain also contain the same receptors for steroid hormones (e.g., androgen and estrogen receptors) that have been described in other vertebrates (see Kelley & Pfaff, 1978; Morrell & Pfaff, 1978 for comparative animal data; for the human brain, see, for example, Kruijver et al. (2001) and Kruijver et al. (2002, 2003)). The cellular and neurochemical "machinery" underlying the expression of sexual behavior in vertebrates is therefore intact in humans. If it had lost its function, one would expect that it should have disappeared or at least evolved in the course of time under the influence of successive mutations. These neural circuits and neurochemical mechanisms sensitive to ovarian and testicular steroids thus presumably still control sexual motivation in humans as they do in other mammals.

It is clearly beyond the scope of this presentation to review the evidence supporting the idea that sexual behavior and in particular sexual motivation is under the control of sex steroids in humans. This conclusion cannot be derived, like in animals, from truly experimental studies in which subjects would be intentionally treated with exogenous steroids and their behavioral responses evaluated. However, even in the absence of such direct evidence, data derived from both correlation analysis and from the treatment of clinical cases clearly highlight the steroid-dependence of human sexual motivation.

Human sexual behavior is complex and its control involves multiple cognitive aspects (experience, social expectations, individual preferences, social conventions, religious and philosophical influences, . . .). Therefore, the mere presence of testosterone in the blood and brain of a man and of estradiol in a woman will not automatically lead to sexual activity. This human complexity must obviously be taken into account, but it does not negate the fact that steroids still have a major impact on sexual motivation. If sexual motivation was entirely the result of learning and a social construction, one could then wonder why no society has ever managed to put its expectations in line with the behavior of individuals. For example, sex outside marriage is discouraged or prohibited in many human societies but remains present even if punished by the death penalty including by methods that could be considered as very discouraging, such as lapidation (stoning).

8.1 SEX STEROIDS AND SEXUAL BEHAVIOR IN HUMANS: CORRELATIVE STUDIES

Sex steroids are not present in constant quantities throughout life. If they play a significant role in the control of sexual activity, one would expect to detect positive correlations between changes in time of the concentrations of steroids and frequency of sexual activity. This is indeed what has been observed in many studies of both men and women.

This correlation is especially observed during puberty. Researchers have analyzed in great detail this period of endocrine instability and demonstrated that the increase in testosterone concentrations is tightly correlated with the sexual behavior of young boys. Higher than average testosterone plasma levels as reflected by salivary concentrations are clearly associated with greater sexual activity (Halpern et al., 1998). This positive correlation was observed both in transversal studies of a large number of individuals at the same age (Udry et al., 1985) and in longitudinal studies that followed a smaller number of subjects during their development (Halpern et al., 1998). These relationships appear to be independent of society and culture in which the boys live and have been for example observed in the Western world as well as in a population living in Zimbabwe (Campbell et al., 2005).

In women, a similar relationship has been identified between the individual values of blood testosterone concentrations and the onset of sexual activity (Halpern et al., 1997). Somewhat unexpectedly, testosterone is indeed involved in the activation of sexual motivation in women, as it is in men (Sherwin & Gelfand, 1987).

Since ovarian activity is cyclical in women, it is also possible to analyze the relationship between hormonal changes and spontaneous behavior (sexual activity, motivation, attractiveness, . . .) in adulthood. Positive correlations have been detected (see LeVay & Valente, 2006, page 142) but many results have admittedly been difficult to reproduce for a variety of reasons ranging from inadequate characterization of the hormonal status of women studied (a single assay of blood concentrations that is not representative of the average hormonal state because of rapid fluctuations of hormone levels, hormonal status inferred from the stage of the menstrual cycle but not confirmed by direct measurements of plasma levels) to interference of non-hormonal factors (daily problems interfering with sexual desire, unrecognized medical problems, . . .).

Some studies suggested the existence of a peri-ovulatory peak of sexual activity in women that would be reminiscent of the estrus in other mammals such as rodents. The interpretation of this peak remains difficult, however, because it could equally result from an increase in sexual motivation of women or from an increase in her attractiveness. Interestingly, a recent study identified periodic changes in the women's attractiveness by quantifying across the menstrual cycle the total amount of tips collected by "lap dancers," i.e., topless girls dancing in night clubs on the knees of customers. Gratuities earned by women during their fertile period (peri-ovulatory) were two times higher than what they earned during their luteal phase or during menstruation. These variations in gains were not observed in women using the contraceptive pill which equalizes the circulating hormone concentrations during most of the cycle (Miller et al., 2007). These data support the idea that sexual attractiveness in women varies during the ovarian cycle as a function of the circulating concentrations of sex steroids. Further research should be undertaken to test the reproducibility of these findings and identify the nature of signals that modulate the attractiveness (verbal, visual, olfactory . . .).

8.2   SEX STEROIDS AND SEXUAL BEHAVIOR IN HUMANS: CLINICAL CASES

Numerous clinical studies have been conducted on different types of patients suffering from varied sexual problems. These studies reinforce the notion that sexual activity in men and women is strongly influenced by sex steroids. For example, in men with low testicular development that produce abnormally low testosterone concentrations (hypogonadism), the decreased sexual activity (low frequency of nocturnal erections, episodes of masturbation and orgasms) can be markedly enhanced by administration of exogenous testosterone (Davidson et al., 1979; Bancroft, 1995; Hajjar et al., 1997; Snyder et al., 2000; Wang et al., 2000). No improvement is, however, observed following administration of testosterone to subjects suffering from lack of sexual desire in which the concentration of circulating testosterone is normal. On the other side, pharmacological manipulations that lower levels of circulating testosterone induce a decrease in libido and various aspects of sexual behavior in humans (Rosler & Witztum, 1998).

Testosterone is also prescribed in combination with estrogen to increase sexual desire in women whose ovaries were removed for medical reasons. There are several controlled studies demonstrating an increase in sexual desire, sexual fantasies and sexual arousal in response to treatment with androgens combined with estrogens. Estrogens alone had in contrast no effect (Sherwin & Gelfand, 1987).

Together these correlational and clinical studies clearly indicate that sexual desire in humans is like in other mammals under the control of sex steroids even if admittedly many other non-hormonal factors modulate this motivation.

8.3   HORMONES ARE "NORMAL" IN HOMOSEXUAL ADULTS

Based on the observation that sex steroids control the activation of sexual behavior in animals and in humans, it was originally hypothesized that modifications in circulating concentrations of these hormones were involved in the control of homosexuality. It is indeed established that, in rodents, for example, testosterone activates the male-typical sexual behavior and estradiol combined with progesterone activates female-typical sexual behavior. It was thus suggested that excessive concentrations of estradiol in some men are responsible for their sexual attraction to other men (a female characteristic) and that vice versa high concentrations of testosterone induce female sexual attraction towards other women (a feature normally observed in men). This old theory is obviously inconsistent with animal studies that have now clearly established that the type of behavior produced by males or females is not conditioned by the type of hormones they are exposed to in adulthood but by the sex of their brain which is determined during ontogenesis (organizing effects of sex steroids, see Chapter 5). It is also impossible to reconcile with the fact that many behavioral effects of testosterone in males are in fact mediated at the cellular level by estrogens locally produced by aromatization of testosterone. Nevertheless, this theory has historically been regarded as plausible and it was therefore experimentally tested.

When accurate assays of sex steroids became available during the 1960s and 1970s, many studies compared the concentrations of circulating sex steroids in men and women in relation to their sexual orientation. All this work led to the very clear conclusion that there are no hormonal differences between homo- and heterosexual men or between homo- and heterosexual women (Meyer-Bahlburg, 1984). Gay men have circulating testosterone concentrations that are perfectly normal and the same applies for the concentrations of estradiol and progesterone in homosexual women. Sexual orientation is thus not linked in humans to an inadequate activation by sex steroids. This conclusion is also in agreement with more recent animal studies showing that a) the type of behavior patterns expressed by male or female rodents is not controlled by the type of steroids they are exposed to but by the sex of their brain, b) the sex of the preferred sexual partner is similarly not affected by steroid action in adulthood but rather depends on the early organizing effects of these steroids and c) circulating testosterone concentrations are not altered in male-oriented ("homosexual") sheep as compared with female-oriented ("heterosexual") sheep.

In conclusion, circulating concentrations of sex steroids are perfectly "normal" in adult gay men and lesbian women. Incidentally, this also explains why all attempts to change sexual orientation through hormone treatments have always failed.

•   •   •   •

CHAPTER 9

The Organizational Role of Steroids on Sexual Orientation: Analysis of Clinical Cases

Besides their role in the activation of human sexual motivation, sex steroids also play a critical role in the sexual differentiation of the genital morphology and of various aspects of behavior. We shall first present a brief overview of the role of steroids in human sexual differentiation before considering how the analysis of clinical cases also suggests that these steroids contribute to determine sexual orientation.

9.1   SEXUAL DIFFERENTIATION IN HUMANS

In the early stages, human embryos are not sexually differentiated and both sexes possess tissues that are at the origin of both the male and female reproductive tracts. These early fetuses have a genital tubercle that can form either a penis or a clitoris and the genital folds that will form the labia (lips) of the vulva in women or the scrotum in men. They also have two sets of ducts, the Wolffian and Müllerian ducts that will later develop into the oviduct and uterus or the epididymis, respectively.

In male embryos, the undifferentiated gonad will rapidly develop into a testis under the influence of the SRY gene located on the Y chromosome (Berta et al., 1990; Sinclair et al., 1990) whereas in the absence of this gene (and of a Y chromosome) female embryos, under the influence of other additional signals, will develop an ovary. The hormones secreted by the testes, mainly testosterone, will then masculinize the morphological, physiological and behavioral traits of the subject as observed in other mammalian species (Phoenix et al., 1959; Jost et al., 1973; Jost, 1985; De Vries et al., 2002).

Testosterone produced by Leydig cells of the fetal testes causes the differentiation of Wolffian ducts into epididymis, vas deferens and seminal vesicles. In parallel, Sertoli cells produce a peptide hormone, the anti-Müllerian hormone or AMH, which as its name suggests, induces rapid and complete regression of the Müllerian ducts. In the female embryo, ovaries secrete little or no hormones. Müllerian ducts spontaneously develop while Wolffian ducts regress completely.

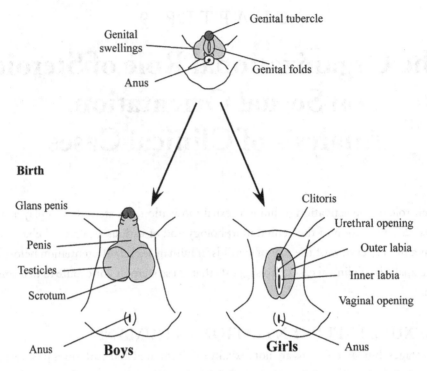

FIGURE 11: Schematic description of the sexual differentiation of the external genital structures in humans. Undifferentiated genital structures are present during the first weeks after conception. Under the influence of testosterone and its metabolite 5α-dihydrotestosterone (DHT), the undifferentiated structures will develop into a male pattern. In the absence of these hormones, the adult female genitalia will develop (see text).

Depending on whether testosterone is present or not, the external genitalia will similarly differentiate into a male or female pattern (Figure 11).

Testosterone and more specifically its androgenic metabolite, 5α-dihydrotestosterone (DHT) produced by the enzyme 5α-reductase, induces the development of male external genitals by fusion of the folds to form a scrotum and the development of the genital tubercle in a penis. In the absence of DHT, the genital folds do not fuse and form the vaginal lips while the genital tubercle develops only very little and turns into a clitoris. Thus, in humans as in other mammals, no hormone production is required for normal development of the female reproductive system that is generally regarded as "default" or neutral sex (see, however, Bakker et al., 2002a; Bakker et al., 2006; Brock &

Bakker, 2011) for a more detailed evaluation of this idea). Testosterone and its metabolite DHT are, in contrast, required to impose masculine development.

These principles that control the sexual differentiation of the reproductive morphology in animals and men are also involved in the emergence of many differences affecting sexual behavior. This notion is clearly established in multiple animal species, where male-typical behavior develops by masculinization under the influence of embryonic testosterone. Female-typical behavior develops in the absence of hormones and is lost in males following embryonic exposure to testosterone (defeminization)(Goy & McEwen, 1980; McCarthy & Ball, 2008). In humans, it is obviously more difficult to demonstrate this relationship between adult behavior and exposure to specific endocrine conditions during the embryonic life but a substantial amount of circumstantial evidence nevertheless indicates that this concept can also be applied in our species. We recently reviewed evidence supporting this idea as far as non-reproductive behaviors are concerned (Balthazart, 2011) and additional reviews on this topic can be found in other books published during the last decade (Baron-Cohen, 2004; Hines, 2004; Baron-Cohen, 2006). We shall specifically focus during the next sections on the organization by prenatal hormones of sexual orientation.

9.2   EMBRYONIC HORMONES AND SEXUAL ORIENTATION

Knowing that in animals sexual differentiation of reproductive behavior and of its orientation is controlled by the action of embryonic steroids (high testosterone levels masculinize and possibly defeminize males in mammals), researchers, and among the first of them the German endocrinologist, Günter Dörner, proposed a theory of homosexuality based on hormonal embryonic imprinting (Dörner, 1969). According to this theory, human fetuses destined to become gay in adulthood would be exposed to an atypical sexual differentiation due namely to the presence of abnormally low testosterone concentration in boys or a too high level of this hormone in girls (see also Chapter 7 on this topic).

It has been argued that such a theory is impossible because it would imply that homosexual men have feminized (or at least less masculine) genital structures whereas lesbians should have experienced some masculinization of genital structures due to the excess of androgens. Although detailed attention has been given to this possibility, no data supporting such differences in genital morphology have ever been collected.

It must be recalled however that genital structures differentiate very early during embryonic life, well before the brain (supporting behavior). Therefore, it is possible that the hormonal imprinting taking place at these two time periods is substantially different. A male embryo could be exposed to high (male-typical) concentrations of testosterone during the first 3 months of gestation (and have perfectly masculine genitalia) but then experience a drop in testosterone concentration toward the end of embryonic life and thus fail to develop a male typical sexual attraction.

Alternatively, the differentiation of the brain and sexual orientation could diverge from genital morphology because steroid action in the brain may be affected while remaining normal in the genital skin. Remember that to produce its effects, testosterone must often be metabolized in its target structures (aromatized in the brain, 5α-reduced in the genital area), bind to specific receptors and then activate complex intracellular signaling cascades eventually leading to changes in protein synthesis and/or neural activity. Any aspect of this complex suite of actions could be affected in the brain but not in the periphery, leading also to a discrepancy between somatic sex differentiation and sexual orientation.

Although this theory remains speculative (and is likely to remain unproven due to the logistic difficulties mentioned before), there are two types of evidence strongly suggesting that it contributes substantially to the determinism of sexual orientation in humans. First, a number of clinical conditions are associated with well-defined endocrine changes during embryonic life and multiple studies have documented an increased incidence of homosexuality in the affected subjects, as predicted by the organizational theory. Secondly, many studies have demonstrated that in gays and lesbians, a number of behavioral, physiological and even morphological features that are known to be sexually differentiated are significantly different from what they are in heterosexual subjects of the same sex. Since in many cases, the sexual differentiation of these features is known to be controlled by embryonic sex steroids, at least in animals, these atypical features suggest that, on average, these homosexual subjects were not fully organized by sex steroid action during their early life in a manner that corresponds to their gonadal sex.

These two types of evidence supporting the idea of organizational effects of steroids on sexual orientation are summarized in the next three chapters. We shall begin by considering the increased incidence of homosexual orientation associated with endocrine problems experienced during embryonic life. Four types of clinical conditions are useful in this context. The first two specifically affect the girls, the following two concern the boys.

9.3   ANALYSIS OF CLINICAL CASES: GIRLS

9.3.1   Adrenal Hyperplasia in Women

Congenital hyperplasia (excessive development) of the adrenal glands (CAH) is a genetic disorder that affects one of the enzymes involved in the synthesis of cortisol (often 17α- or 21-hydroxylase) in the adrenal glands. This indirectly results in the hypersecretion of androgens (compounds similar to testosterone) in the affected embryos (Hines, 2003; MacLaughlin & Donahoe, 2004). As a result, affected female individuals display a more or less profound masculinization of the external genital structures. In the most severely affected subjects, the genital tubercle has developed into a fully formed penis with a size equal to that seen in male babies. In addition, there is a more or less complete fusion of the genital lips that form a scrotum. There are, of course, no testicles and the internal reproductive organs are those of normal girls.

This more or less complete masculinization of external genital structures in CAH girls is usually detected at birth. These children are treated during their entire life by administration of exogenous glucocorticoids that will also reduce the inadequate production of androgens through their feedback action on the brain and pituitary gland. In addition, the masculinized external genital structures are often "corrected" surgically, removing part of the penis to reproduce a clitoris and incising the genital folds to reopen the vaginal opening. These children are then raised as girls.

These girls display during childhood a masculinization of certain character traits such as the type of toys used or freely made drawings (Berenbaum & Hines, 1992; Hines & Kaufman, 1994; Berenbaum & Snyder, 1995; Berenbaum et al., 2000; Iijima et al., 2001; Hines et al., 2003; Hines et al., 2004). They also show a high male-typical level of physical activity (see Hines, 2004; 2011 for review). They provide a unique source of information on the potential role of embryonic androgens in the control of sexual orientation since they were exposed to an excess of androgens during embryonic life but then raised, as far as we can tell, as girls. Several researchers have therefore studied whether there was a change in the incidence of homosexuality in this population of girls and generally, a positive response has been given to this question.

Multiple studies have indeed detected in CAH women an increase of the probability of homo- or bisexual erotic imagery and of the commitment or desire to engage in a homosexual relationships in comparison to a population of control females or to unaffected sisters of these androgenized women (Money et al., 1984; Dittmann et al., 1992; Zucker et al., 1996). While in control populations, the incidence of female homosexuality is estimated to be around 10–12% (according to the Kinsey report (Kinsey et al., 1953), more recent studies show even lower percentages below 5%, see Mosher et al., 2005), a study by John Money and his colleagues reported for example an incidence of homo-or bi-sexuality of 37% in CAH women (Money et al., 1984). More recent studies sometimes based on much larger sample sizes have produced similar figures (Hines, 2006; Meyer-Bahlburg, 2008; Meyer-Bahlburg et al., 2008). The study of Meyer-Bahlburg et al. (2008) also showed that the degree of virilization in various forms of CAH correlates with the increased incidence of bi- or homosexuality (Figure 12). This paper also reviewed and presented a summary of 18 studies on the subject in which most studies showed an increased incidence of homo- or bi-sexuality in CAH girls and thus a reduction of conventional heterosexual orientation.

The theoretical interest of this medical condition is that postnatal education of the subjects (as a girl in theory) is expected to act in the opposite direction of the prenatal hormonal influence (masculinization). The masculinization of a behavioral trait (female homosexuality, i.e., sexual attraction for women, a trait normally male) should thus in theory result from prenatal hormonal effects.

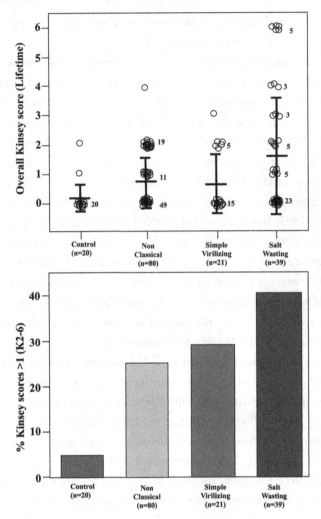

FIGURE 12: Effect of prenatal androgenization linked to the syndrome of congenital adrenal hyperplasia (CAH) on sexual orientation. **A.** Individual data are reported for control (C) subjects and for three subtypes of CAH producing increasing levels of virilization from the non-classical (NC) to the simple virilizing (SV) to the salt-wasting (SW) type. Sexual orientation is reported based on the 7 point Kinsey scale ranging from 0 (K0, completely heterosexual) to 6 (K6, completely homosexual). When multiple data points are present at the same score, their number is indicated on the right side of the cluster. **B.** Percentages of bi- or homosexual subjects expressed as the ratio of the numbers of subjects rating between K2 (largely heterosexual but also distinctly homosexual) and K6 to the total number of subjects (K0-1 + K2-6). This percentage increases with the severity of the CAH phenotype (from C to NC to SV to SW). Drawn from data in Meyer-Bahlburg et al. (2008).

It has been suggested that this change of sexual orientation could be caused indirectly by the masculinization of the external genital structures. Although these structures are surgically corrected at birth, they still do not have an optimal shape and may therefore allow little or no penetrative heterosexual relationships. The data available to evaluate the specific contribution of this factor are still limited but it is clearly involved. Further studies are needed. It is also possible, theoretically, that the increase in non-heterosexual behavior and attraction in CAH women is induced indirectly by reactions of the parents of the girl with masculinized genital structures. Parents might not educate their daughter in the same way as an unaffected girl. However, if anything, parents should try to promote feminine aspects of these girls knowing that the little girl is genetically female but was partially masculinized during gestation. One could then imagine that the efforts of parents systematically produces an effect opposite to the intended effect, which would be paradoxical. At this stage, the most logical interpretation of the change of sexual orientation observed is therefore that it was induced by the exposure to prenatal androgens. More subtle interactions between the prenatal endocrine condition and postnatal environment could however be implicated in the determination of sexual orientation and will be considered in Chapter 13.

Note however that adrenal hyperplasia is a rare disease that cannot be invoked as a general determinant of sexual orientation of all lesbians. In addition, the effect of prenatal androgens in CAH girls, even if it is very significant in statistical terms, has limited amplitude. Only 30–40% of the population of the affected girls are not strictly heterosexual which leaves 60–70% of the CAH girls unaffected in terms of sexual orientation.

The reasons that potentially limit the effects of embryonic androgens on sexual orientation might include the existence of limited periods of sensitivity to androgens that overlap only partly with the periods when these steroids are present in high concentrations (problem of time-response) or the fact that sexual orientation is only in part controlled by prenatal androgens and that action of other factors is required. Data that could discriminate between these interpretations are currently not available.

9.3.2   Treatment of Pregnant Mothers with DES

Between 1939 and 1960, about two million pregnant women were treated with the synthetic estrogen diethylstilboestrol (DES) in Europe and the United States to prevent unwanted miscarriages. This treatment was not only ineffective but DES caused a slightly elevated risk of cervical cancer in girls exposed during their embryonic life and was also associated with a masculinization of reproductive function and behavior. These treatments were discontinued as soon as scientists became aware of their consequences but people unfortunately exposed to DES action have provided another important source of information about the long-term behavioral effects of embryonic steroids on sexual orientation.

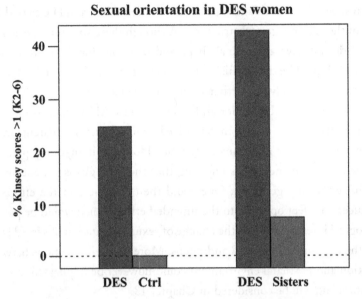

FIGURE 13: Effect of prenatal treatment with diethylstilboestrol (DES) on sexual orientation in women. A significant increase in the percentage of individuals that are bi- or homosexual (Kinsey scores between 2 and 6) is observed in treated women as compared with control subjects. The same increase is also observed in the comparison of the sub-population of women exposed to DES and their non-exposed sister. Drawn from data in Ehrhardt et al. (1985).

Several studies have indeed demonstrated that girls exposed before birth to DES are more likely to be bi- or homosexual (Ehrhardt et al., 1985; Meyer-Bahlburg et al., 1995; Swaab, 2007). In the first of these studies, 24% of 30 women exposed to DES during their embryonic life were classified on the Kinsey scale with scores between 2 and 6 (bi- or homosexuality), while none of the 30 control women (matched on as many variables as possible) were scoring in this range (Figure 13). Furthermore, twelve of these DES-exposed girls had a control non-exposed sister and comparison of these two closely matched populations similarly indicated an increased incidence of bi- or homosexuality in subjects exposed to DES (Swaab, 2007) (5/12 vs. 1/12).

An independent study similarly showed an increase of most measures of homosexual orientation (ideas associated with masturbation, dreams, attraction, relationships) in a population of DES women as compared to their controls as well as in a sub-population of sisters where only one of the two women had been exposed to DES before birth (Meyer-Bahlburg et al., 1995). In contrast, a more recent study suggests that girls exposed to DES before birth have a slightly lower probability of having sex with a same-sex partner as control subjects (Titus-Ernstoff et al., 2003). The origin of this discrepancy is unknown but could be related to the fact that this study considers only the actual sexual activity, whereas previous studies also analyzed other aspects of sexuality (fantasies, dreams) and used a graded scale (Kinsey scale in 7 points) rather than a dichotomous classification (presence/absence of a homosexual relationship).

If the effect of DES is real, which will be difficult to confirm given that this treatment has now been abandoned long ago and exposed subjects are becoming less and less available, this would indicate that estrogens as well as androgens (testosterone) are able to masculinize sexual orientation. This is quite plausible since in rodents, testosterone exerts many of its effects on sexual differentiation after conversion into estradiol by aromatase in the brain (see Chapters 5–6). It should however be noted that studies conducted in rhesus monkeys and a number of data from the clinic indicate that human sexual differentiation might be under the direct influence of (non-aromatized) androgens, which would make the effects of DES more difficult to interpret.

9.4 ANALYSIS OF CLINICAL CASES: BOYS

9.4.1 5α-Reductase Deficiency

As already mentioned, testosterone must be converted into the more potent androgen 5α-dihydrotestosterone (DHT) in the skin of the genital area in order to promote the development of the genital tubercle into a penis and the fusion of genital folds to form a scrotum. The conversion of testosterone into DHT is catalyzed by an enzyme called 5α-reductase which is expressed at high levels in the genital tubercle and skin folds. Imperato-McGinley and colleagues identified, in the Dominican Republic, a mutation of the 5α-reductase enzyme, which makes it functionally unable to produce DHT (Imperato-McGinley, 1994; Imperato-McGinley & Zhu, 2002). This recessive mutation has also been discovered since in Papua New Guinea, another island where inbreeding is important.

Male (XY) embryos affected by this mutation are born with external genital structures that are only partially masculinized. Their apparently normal testes secrete the anti-Mullerian hormone that induces regression of structures derived from Müllerian ducts (fallopian tubes and uterus). Testosterone also promotes the development of male-typical internal sexual structures (epididymis, seminal vesicle, prostate) but, externally, at the level of external genitalia, the genital tubercle is not developing and remains about the size of a clitoris. There is no fusion of genital folds: at birth the two labia surround the opening of a blind vagina and consequently there is no scrotum and testes remain in the inguinal canal.

Around puberty, the dramatic increase in circulating testosterone concentrations will however induce a partial masculinization of the genital structures: the skin of the genital folds wrinkles and darkens so it looks more like a scrotum, the testicles will continue their descent to end in the genital folds and the genital tubercle will develop slightly to resemble a small penis. The opening of the urethra however remains at the base of the "penis" a medical condition called hypospadias. Testosterone in pubertal males also affects other aspects of body morphology; it increases muscle size and lowers the tone of the voice. Superficially, it seems that at puberty these individuals have changed sex, at least partially.

Boys affected by this change are essentially raised as girls after having been exposed to androgens (testosterone but not DHT) and probably also estrogens during early life. They provide therefore another test of the relative power of the potentially opposite effects on sexual orientation of the embryonic endocrine environment versus education.

During their childhood, these XY subjects always adopt a female gender identity but after the testosterone-driven (partial) masculinization of their genital morphology at puberty, they usually change their sexual identity in adulthood to live as men. This change of identity and gender role has even been observed in some individuals who had been married at a very young age to a man as a girl and who remarried with a woman after puberty. In terms of sexual orientation, these adults usually show a male-typical attraction to women, which is thus in agreement more with the hormonal impregnation during early life than with the postnatal education and life as a girl.

These studies have been criticized because 5α-reductase-deficient children are usually identified at birth. They have a specific name, "Guevedoche," in the Dominican Republic which in Spanish means "eggs, i.e. testes at twelve years" or "Kwolu-aatmwol" in Papua, which means "female thing transforming into male thing" and one could imagine that they are not really (fully) raised as girls, but rather as inter-sex since their family knows from the start that masculinization is going to occur at puberty (Rubin et al., 1981). It was also suggested that the sex change occurring at puberty is motivated by the privileges granted to men (especially in terms of independence, ability to undertake studies . . .) in the male-dominated societies where this phenomenon is observed (Rubin et al., 1981).

Imperato-McGinley and her colleagues showed, however, that in a group of 18 affected subjects who had completely female external genitalia at birth and had therefore been brought up in the firm belief that they were small girls, 17 adopted after puberty a male sexual identity and were sexually attracted to women (Imperato-McGinley et al., 1991). It is also important to note that the same change in sexual identity and orientation after puberty has been observed in 5α-reductase deficient subjects of Papua New Guinea, i.e., in a society that less easily accepts a change of gender identity and sexual orientation (Imperato-McGinley et al., 1991). Because it is largely accepted that gender identity and sexual orientation become definitely fixed during childhood and are later very difficult if not impossible to change, these data argue for a role of the early endocrine environment as opposed to postnatal education and socialization in the determination of these sexually differentiated characteristics.

Many questions, of course, remain open regarding these studies. If we imagine that the change of identity and sexual orientation observed at puberty is the result of effects of androgens during fetal life (organizing effects) and puberty (activating effects) that are able to counteract almost completely the social experience and education, we are then led to believe that the action of testosterone on gender identity and sexual orientation is due to the hormone itself or its estrogenic metabolites and that conversion to DHT, the active metabolite at the peripheral level is irrelevant for the behavioral responses. Also the idea that these children are truly and fully raised as girls will always be impossible to prove. The fact remains that, as in the case of adrenal hyperplasia, if education has a role, it always leads to the opposite effect of what is intended which would be ironic to say the least.

9.4.2   Cloacal Exstrophy

Cloacal exstrophy is a rare complex genito-urinary malformation occurring during embryonic development that results in the birth of XY males who, in addition to various malformations of the pelvis, have a poorly developed or no penis. It is not an endocrine disease in that the testes are apparently normal both morphologically and functionally. In most cases, the XY individuals are assigned at birth a female sex both at the legal and social levels. They also receive corrective surgery including removal of the testes and vaginoplasty.

Several recent studies have followed the psychosexual development of affected subjects and, surprisingly, it was observed that in a large number of cases (sometimes up to 55%, 8 of 14 XY individuals) subjects adopt as adult a male identity and gender role (Reiner & Gearhart, 2004). Another study reported that among 19 XY subjects assigned a female identity before the age of 12 years, 10 were living as female when interviewed at a later age but 9 had switched to a male gender identity (Meyer-Bahlburg, 2005). A clear masculinization of preferred games played by these XY children reared as girls was also present (preference for physical sports [football] or aggressive sports [karate]) (Reiner & Gearhart, 2004). Information concerning the sexual orientation of these subjects is sparse, but in many cases, a typical male sexual orientation (attraction to women) was also observed.

These data again suggest that hormonal imprinting of the embryo by androgens could play a critical role in the development of sexually differentiated characteristics such as gender identity and sexual orientation. Importantly, these masculine features develop in the absence of pubertal action of testosterone since most subjects were castrated soon after birth. Prenatal or immediately postnatal androgens would thus explain the high incidence of changes in gender identity (and sexual orientation?) observed later in life in subjects exposed *in utero* to androgens, despite the fact that they were subsequently raised as girls. This change however does not concern all individuals. If this type of "sex change" is consistent with a role of prenatal androgens, it also indicates that androgens are probably not the only determinant. We will come back to this idea.

9.5 OTHER ACCIDENTAL CHANGES IN THE EMBRYONIC HORMONAL MILIEU

The four conditions affecting sexual orientation described above contribute to the explanation of this feature because the assumed role of hormones goes against the supposed role of education. It is therefore possible to differentiate to some extent the impact of these two influences. There are also other endocrine changes that affect sexual orientation and put it into conflict with the genetic sex of the affected individuals, but these clinical cases are less interesting for the purpose of this work because in these cases, the effect of education is consistent with the effect of the endocrine disease and therefore one cannot easily distinguish their respective roles.

For example, males (XY) affected by a complete androgen insensitivity due to mutations of the androgen receptor are born with a completely female genital morphology. They are then raised as girls and in general adopt a female gender identity and a female sexual orientation. Detailed monitoring of a group of these patients indicated that they were very satisfied with their sexual identity and their sex life in general (Wisniewski et al., 2000). However, it is impossible in these cases to know whether the gender identity and sexual orientation as women are the result of the post-natal education or of the lack of action of testosterone during embryonic life. These cases show however that gender identity and sexual orientation are not directly linked (and thus determined) by the genetic sex.

Similarly, various chromosomal abnormalities (XXY or XYY boys and XO girls) provide little information for understanding the mechanisms of sexual orientation since the missing or supernumerary chromosomes contain many genes whose absence or presence could directly or indirectly influence sexual orientation. It is thus impossible in these cases to link sexual orientation with any certainty to a specific genetic or hormonal factor.

9.6 IN CONCLUSION

We have in this chapter reviewed a number of medical conditions that are associated with increases in the incidence of homosexuality. In these conditions, a sexual orientation is often adopted that is in opposition to the sex assigned at birth, but consistent with the type of presumed hormonal exposure during intrauterine life (attraction to men if absence of androgens and to women if presence of androgens during embryogenesis). Most of the effects described were reproduced in independent studies on populations of unrelated subjects, which adds to their credibility. It must, however, be accepted that, occasionally, these effects could not be replicated in other studies. This is not necessarily surprising given the complexity of these problems, including the selection of subjects and their controls and the difficulties associated with an accurate determination of gender identity and sexual orientation in subjects whose education and hormonal history were potentially disturbed.

Taken together, the studies nevertheless strongly suggest that hormones acting during embryonic life substantially contribute to the determination of sexual orientation as they do in a variety of animal species. The most convincing medical cases suggesting a role of embryonic hormones on sexual orientation are those in which the effects of prenatal hormones and of education are supposed to pull in opposite directions (e.g., congenital adrenal hyperplasia, prenatal DES exposure, 5α-reductase deficiency and cloacal exstrophy). However, because only a fraction of subjects (often 20–40%) were affected in all studies (except in the androgen insensitivity syndrome but here it is impossible to separate the role of embryonic hormones and of education), the data are also incompatible with a full determination of sexual orientation by prenatal androgens. One can thus imagine that the prenatal hormonal environment only predisposes to a sexual orientation (and identity) but does not fully determine them. Alternatively, it is also possible that the endocrine disorders that are studied have a variable amplitude and duration so that they do not affect equally all individuals because they do not occur at the critical moment or do not reach a critical amplitude to modulate aspects of sexual differentiation of the brain responsible for sexual orientation. We will come back to these different possibilities at the end of this presentation.

· · · ·

CHAPTER 10

Sex Differences Not Related to Sexual Activity Suggest an Atypical Differentiation in Homosexual Subjects

There is also a large amount of correlative data strongly suggesting that sexual orientation in humans is influenced, like sexual partner preference in many animal species, by the early hormonal imprinting. The most direct approach to the study of this question would obviously be to measure hormones, in particular androgens, at multiple time points during fetal life in a large number of subjects and then correlate these data with the sexual orientation of the subjects 20 or plus years later. This approach is obviously challenging from a logistic point of view (cost, duration of the study, small proportion of subjects that will later become homosexual as adults) but also impossible for ethical reasons. Blood collection in embryos is associated with a small but nevertheless significant risk of abortion that is taken only when required by serious medical reasons, but not to make a scientific study.

Researchers must therefore rely on indirect evidence of what could have been the hormonal embryonic milieu of homosexual individuals. They have identified morphological, physiological and behavioral traits that are different between men and women and in some cases it has been demonstrated that these differences result from a differential exposure to sex steroid of male and female embryos. For many of these sexually differentiated traits, definitive evidence is available in animals to demonstrate that they depend on embryonic steroids but evidence is only suggested by correlative data in humans. Many studies have evaluated these characteristics in a comparative manner in populations of homo- and heterosexual subjects. If homosexual subjects of one sex express a given trait in a manner that is characteristic of the other sex and if the prenatal endocrine environment affects this trait, it can indeed be suspected that these subjects were exposed to atypical hormonal conditions during embryonic development. Many positive results have been gathered in this type of study and are summarized in this chapter.

10.1 COGNITIVE AND BEHAVIORAL DIFFERENCES NOT DIRECTLY RELATED TO SEX

There are many differences in average cognitive abilities between men and women. Although most of these differences have limited amplitude, they are nevertheless reproducible. Education clearly plays a major role in the genesis of these differences but some of them appear to result also in part from sex differences in hormone concentrations during early life, i.e., from organizing effects of steroids (Collaer & Hines, 1995; see also Baron-Cohen, 2004 for a more detailed discussion of this interaction between biological and cultural factors).

If homosexuality is due, at least in part, to an atypical embryonic hormonal milieu, these sexually differentiated characteristics should logically be different in homosexual compared to heterosexual subjects within a same sex. Several studies indicate that this might be the case.

10.1.1 Tests for which Men are on Average Better than Women

10.1.1.1 Visuo-Spatial Tasks. Homosexual men have mean performances inferior to those of heterosexuals in the execution of several tasks, including visuo-spatial mental rotation tasks, evaluation of the orientation of a straight line and aiming at a specific target (e.g., Hall & Kimura, 1995; Neave et al., 1999; Rahman & Wilson, 2003b). Their performances are similar to those of women or intermediate between men and women. In contrast, several studies have failed to identify such differences between homo- and heterosexual males sometimes for the same tasks (Gladue et al., 1990; Tutle & Pillard, 1991).

Studies conducted in lesbian women who should according to the hormonal theory display masculinized performance have usually failed to identify clear differences. One study showed however that homosexual women perform mental rotations slightly better than heterosexual women but the difference only affects the speed of execution of the task, not its accuracy (Rahman & Wilson, 2003b).

10.1.1.2 Aggression. In general, men are more aggressive than women and two separate studies have indicated that gay men were less physically aggressive than heterosexual men (Ellis et al., 1990; Gladue & Bailey, 1995). No difference in aggression has been identified between lesbians and heterosexual women (Gladue & Bailey, 1995) perhaps because the basal level is so low that a small increase remains unnoticed.

10.1.2 Tests for Which Women are on Average Better than Men

10.1.2.1 Remembering the Location of Objects. A study based on a large number of subjects showed that gay men more efficiently localize objects than heterosexual subjects, which is consistent with the idea that this feature has not been fully masculinized in homosexuals. However, no difference was observed in this study between homo- and heterosexual women (Rahman et al., 2003b).

10.1.2.2 Verbal Fluency.   A study published in 1991 indicated that homosexual men have a higher verbal performance than heterosexuals which would agree with the idea that they are more feminine from this point of view (McCormick & Witelson, 1991). Two subsequent studies did not confirm this difference (Gladue et al., 1990). A recent analysis based on a large number of subjects reported that both gay men and women have performances in verbal fluency tests, which are atypical for their gender (Rahman et al., 2003b). In one of these tests, gay men were better than all other groups (heterosexual men and women regardless of their orientation). In another test, gay men and heterosexual women were better than lesbians and heterosexual men. Overall these data suggest that for these features gay men are more similar to women while lesbians are more like men.

10.1.3 Features Not Usually Differentiated

10.1.3.1 Manual Lateralization.   Several studies have shown that both gay men and lesbians tend to be either left-handed or ambidextrous (i.e., non right-handed) more frequently than heterosexual individuals (Lalumiere et al., 2000; Lippa, 2003b). This feature is not generally regarded as sexually differentiated in the heterosexual population (similar proportions of right- or left-handed or ambidextrous subjects among men as among women; Lippa, 2003b). The existence of a predominant development of the right side of the body among men while the left side would be favored in women has however been suggested (Baron-Cohen, 2004).

A recent meta-analysis of five independent studies also confirms the existence of a higher probability among homosexual men to be non-right-handed (Blanchard & Lippa, 2007). This increased frequency of non-right handedness does not necessarily fit with the idea that homosexuality reflects an abnormal prenatal hormonal differentiation. However, it is interesting from the point of view of the ontogenesis of homosexuality because the preferential use of one hand over the other is an individual trait observable before birth (Hepper et al., 1991) even if a reversal of lateralization is occasionally possible after a trauma (e.g., difficult birth). The higher incidence of non-right handedness among homosexuals of both sexes would thus be consistent with a prenatal determination of this orientation.

These differences in cognitive abilities related to homosexuality are, of course, consistent with the idea that an atypical embryonic hormonal milieu may have influenced in parallel the two types of variables. However, the hormonal contribution to these cognitive traits is probably limited (the role of the postnatal environment is important here) and many differences have a low reproducibility. The cognitive differences could also be a secondary consequence of the homosexual orientation.

Therefore, these cognitive changes observed in homosexuals support only moderately the hormonal theory of homosexuality. Morphological and physiological differences associated with sexual orientation provide however a much stronger evidence.

10.2 PERIPHERAL MORPHOLOGICAL DIFFERENCES

A good number of morphological traits are also different between men and women and are determined in part by embryonic hormones. Some of these traits are modified in homosexuals suggesting again that they were exposed *in utero* to abnormally high or low concentrations of sex steroids. Two morphological characteristics have been studied from this point of view.

10.2.1 The Length of the Index and Ring Fingers

The length of fingers is not the same for males and females. As an average, in most women, the index (finger 2 or D2) is almost as long as the ring finger (finger 4 or D4) so that the ratio of length D2: D4 is very close to 1 (mean = 0.973). In contrast, in males, the index is shorter than the ring finger and the D2: D4 is smaller (around 0.955) (see Figure 14).

This sex difference in the relative length of fingers is also observed in various mammalian and avian species (Brown et al., 2002a; Romano et al., 2005) and it is in these species significantly affected by early treatments with androgens (Lutchmaya et al., 2004; Romano et al., 2005; Manning et al., 2006). This sex difference in the relative length of fingers in humans is thus probably reflecting differences in the embryonic androgen concentrations (male > females; see Breedlove, 2010 for review). It is well established that testosterone modulates the growth of long bones in the human embryo as it does in other vertebrates. This notion is supported by several studies indicating that girls exposed to high androgen levels due to congenital hyperplasia of the adrenal glands (CAH, see previous chapter) have a masculinized D2:D4 ratio, i.e., a smaller ratio than in "normal" women (Brown et al., 2002c). One recent study also showed that the D2:D4 ratio is significantly higher in XY subjects affected by the complete androgen insensitivity syndrome than in control subjects (Berenbaum et al., 2009).

In agreement with the theory that assigns a role to prenatal androgens in the development of homosexuality, several independent studies showed that the D2:D4 ratio is significantly smaller (and therefore more similar to the ratio observed in men) in lesbians than in heterosexual women (McFadden & Shubel, 2002; Rahman & Wilson, 2003c; Kraemer *et al.*, 2006). Interestingly, only lesbians who identify as masculine (known as "Butch") had this decreased D2:D4 ratio (Brown et al., 2002b). To my knowledge, one single study failed to replicate the relationship between the lower D2:D4 ratio and sexual orientation in women (Lippa, 2003a).

Another study showed that among monozygotic twins (true twins) who have a discordant sexual orientation (one is homosexual and the other not), the lesbian sister had a smaller D2:D4 ratio (more masculine) compared with the heterosexual sister (Hall & Love, 2003). This work also

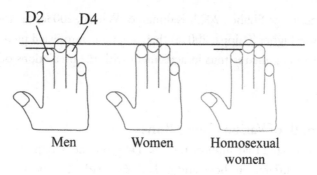

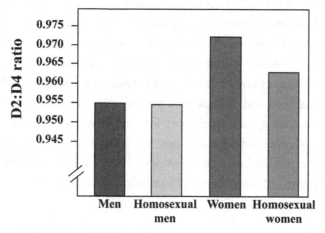

FIGURE 14: Ratio of lengths of the index (D2) and ring finger (D4) and sexual orientation. The ratio D2:D4 is greater in women than in men, probably due to in utero exposure to testosterone of male embryos. It is close to the male level in homosexual women, suggesting that they have been exposed to abnormally high levels of testosterone during a part of their embryonic life. This ratio is not consistently changed in gay men as compared to heterosexual men.

suggests that variations of the D2:D4 ratio is independent of genes since the study is here on twins who have an identical genetic heritage. A more recent study based on a larger number of subjects, however, indicates a genetic contribution in conjunction with a contribution of the prenatal environment in the determination of the D2:D4 ratio (Gobrogge et al., 2008). If testosterone is a key factor that determines this ratio, twins must then be exposed to different concentrations of testosterone, which has so far not been experimentally tested. However, it is also possible that twins exposed to a same concentration of steroids, are differentially sensitive to the steroid.

Similar studies were conducted in men with the expectation that homosexuality would be associated with a more feminine (higher) D2: D4 ratio. Conflicting results were obtained in this work. Either the D2:D4 ratio was lower (more masculine) in homosexual subjects (opposite to the prediction) (McFadden & Shubel, 2002; Rahman & Wilson, 2003c) or it was identical (Williams et al., 2000) or it was higher ((Lippa, 2003a) than among heterosexual men. Only one of these four studies conducted among men is thus in agreement with the predictions of the prenatal hormonal theory.

10.2.2 The Length of Various Long Bones

Like the bones of the hand that support the D2: D4 ratio, the length of the long bones of the arm and leg also become different in boys and girls during early life and may reflect the influence of the sex-specific embryonic hormonal environment (androgenic and estrogenic steroids). One study consequently measured the length of various long bones in men and woman and assessed potential changes in the relationships between these lengths in homosexual individuals (Martin & Nguyen, 2004). It was shown that the length of these bones that become sexually differentiated during infancy was significantly different between homo- and heterosexual individuals while the bones that become different between men and women after puberty were not modified according to sexual orientation. Subjects who have a sexual preference for men (heterosexual women and homosexual men) have a smaller growth of the bones of the arms, legs and hands than subjects who have a sexual preference for women (heterosexual men and homosexual women). These differences were particularly noticeable in the analysis of the ratio of the hand width to hand length (see Figure 15). These

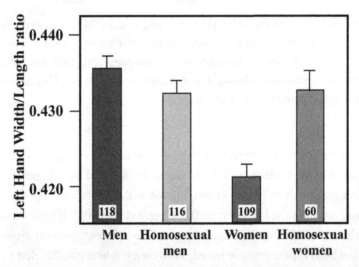

FIGURE 15: Relationship of sex and sexual orientation to the ratio of length to width in the left hand. Data are means ±SEM. The number of subject is indicated in each bar. Redrawn from data in Martin & Nguyen, 2004.

data again support the idea that gay men experienced a reduced exposure to sex steroids during development, and conversely that homosexual women were exposed to sex steroids concentrations higher than heterosexual women. Alternatively, the differences observed may reflect differences in the sensitivity to these hormones rather than their plasma concentrations.

Although the reproducibility of these findings has not been tested to date, it should be noted that before this large study based on nearly 400 subjects (Martin & Nguyen, 2004), other less systematic research had already identified in males differences in ratios of the lengths of various parts of the body associated with sexual orientation (see references in Martin & Nguyen, 2004). It is therefore quite likely that the results of this study are indeed reproducible.

10.3 PHYSIOLOGICAL DIFFERENCES
10.3.1 Oto-Acoustic Emissions

The inner ear, in addition to its obvious function in hearing, also emits sounds in the form of barely audible clicks that can be recorded by placing a sensitive microphone in the canal of the ear. The subjects do not usually hear these sounds called oto-acoustic emissions (OAE). They are produced either spontaneously or in response to external short noises (e.g., clicks) and are an indirect consequence of the functioning of the inner ear. Interestingly, women produce OAE that are more frequent and have larger amplitude than in men. These differences are already present during childhood and are the result of the prenatal exposure to androgens in male embryos (McFadden, 2002) (Figure 16).

This notion is largely supported by studies of OAE in animals. Sexually differentiated OAE are indeed found *mutatis mutandis* in both sheep and monkeys (female frequency and amplitude larger than in males). In sheep, prenatal treatment of female embryos with testosterone considerably reduced the magnitude of their OAE (McFadden et al., 2009). OAE were also studied in hyenas, a species where females are strongly masculinized by androgens during fetal and early post-natal life, so that their external genital structures are at first sight not different from those of males (enlarged clitoris similar to a male penis). Correspondingly, the OAE of female hyenas have a magnitude equal or even slightly smaller than in males. If pregnant hyenas are treated with an antiandrogen (a compound that blocks androgens action at their receptor), their offspring will display OAE with an amplitude well above normal (McFadden, 2008). Finally, in both sheep and hyena, castration in adulthood does not affect the OAE, confirming that the sex difference affecting this characteristic reflects the physiological prenatal hormonal milieu but not activation by steroids in adulthood.

Dennis McFadden and his colleagues at the University of Texas in Austin found that lesbians and bisexual women have spontaneous and click-induced oto-acoustic emissions that are partially masculinized: they emit significantly fewer OAE than heterosexual women, and these OAE have a lower amplitude (McFadden & Pasanen, 1998; 1999; McFadden, 2002). These data are thus

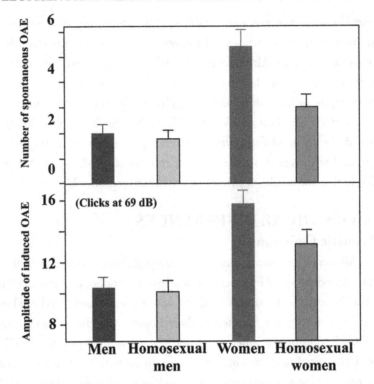

FIGURE 16: The oto-acoustic emissions (OAE) are sexually differentiated and significantly masculinized in homosexual females. The two panels represent the average number of spontaneous OAE (top) and OAE amplitude (bottom) evoked by clicks sounds measured in the right ear of hetero- or homosexual males and females. Redrawn from data in McFadden (2002; 2008).

consistent with the hormonal theory of homosexuality stating that female fetuses destined to become homosexual or bisexual have been exposed to abnormally high levels of androgens. These same researchers also demonstrated that women who had a twin brother and therefore had potentially been exposed to slightly higher levels of testosterone during embryonic life (testosterone produced by the twin would have diffused to the female embryo) had masculinized OAE confirming sensitivity to androgens of the response in humans.

Somewhat surprisingly however, these same researchers have not found any difference between the OAE of homo-and heterosexual males which is obviously contrary to what one might expect from the most simple form of the hormonal theory of homosexuality (homosexual man = exposure during embryonic life to low levels testosterone = higher OAE than in heterosexual men).

10.3.2 Evoked Acoustic Potentials and "Startle Response"

The same group of Texas researchers also showed that evoked acoustic potentials (electrical responses induced in the brain by short auditory stimuli) are also different in men and women and are partly masculinized in homo- or bisexual women compared with heterosexual women. Other characteristics of these acoustic evoked potentials were by contrast hypermasculinized in homosexual men in comparison with heterosexual men (McFadden & Champlin, 2000). This study is again consistent with the idea that lesbians were exposed to abnormally high levels of androgens during fetal life. In contrast and contrary to what could be expected, gay men are for this response equally or even more masculine than heterosexual men. No definitive explanation for this paradoxical observation about gay men has been provided to date.

An independent study in London focused on another physiological response known to be sexually differentiated. Men and women differ for a particular fear response called the alarm or "startle response." Following a loud noise, there is a blink of the eyes but this response is partially inhibited if the loud noise is preceded by a lower noise. This inhibition by the preceding sound ("pre-pulse inhibition") is usually less pronounced among women than among men and the group of Rahman and his colleagues recently showed that inhibition of the blink by a low alarm is stronger among lesbians than among heterosexual women (Rahman et al., 2003a). This response is thus masculinized in agreement with the prenatal hormonal theory of homosexuality. However, here again, authors found no difference between homo- and heterosexual men. No simple explanation is available to explain the fact that gay men do not show a lack of masculinization of these responses as expected if male homosexuality was due to early exposure to low androgen. Some studies have even shown a paradoxical hypermasculinization.

10.3.3 Positive Feedback in Response to Estradiol

In mammals, ovulation is induced in females by a gradual increase in plasma estradiol during the follicular phase of the cycle. When estradiol reaches a critical threshold, it induces by positive feedback the release of a large peak of luteinizing hormone (LH) that ultimately causes ovulation. This LH peak can be induced in female rodents, but not in males, by a single injection of a high dose of estrogens and this neuroendocrine sex difference is organized during the embryonic and perinatal period by the same hormonal mechanisms that differentiate sexual behavior (see Chapter 5). The injection of testosterone in female rat embryos induces the loss of the LH peak in response to estradiol while males castrated immediately after birth are able to produce such a peak.

After suggesting that male homosexuality is the result of inadequate embryonic masculinization, the German researcher Günter Dörner wondered whether the positive feedback of estrogen on LH secretion might be different in homo- and heterosexual men (Dörner, 1969). He showed indeed that gay men react to an injection of estrogen by a significant increase in blood levels of LH. This increase was lower than observed in women but it was still significantly higher than in hetero-sexual men (Dörner, 1972; 1976; 1980).

This work raised a huge controversy, not only because Dörner was openly suggesting to establish preventive androgen treatments of male embryos to prevent the emergence of gay men but also because the positive feedback of estrogens on LH, which is quite evident in rats, is ap-parently not present in monkeys (Baum et al., 1985) which raises the question of its existence in humans. In fact, in women, an injection of estrogens at the appropriate time of the ovarian cycle induces a strong increase of circulating levels of LH, whereas this response is not observed in men. However, the increase observed in women does not have the amplitude and rapid time-course of the pre-ovulatory surge described in rats. It is therefore unclear whether this increase represents a pre-ovulatory peak and more critically for the present discussion whether this increase is sexually differentiated by the action of embryonic steroids.

Many researchers have thus questioned the reality of the effect identified by Dörner and some decided to test its reproducibility. In particular, Gladue and colleagues demonstrated that after a single injection of Premarin, a compound with estrogenic action, women showed 72 to 96 hours later a major increase in LH plasma concentrations (Gladue et al., 1984). This increase was not ob-served in men but again an increase with an intermediate magnitude was observed in homosexual men (see Figure 17).

Although this LH response is similar but not identical to the preovulatory peak observed in rats (see for discussion Baum et al., 1985), these results clearly establish the existence of an endo-crine difference in men according to their sexual orientation. This difference may concern a neuro-endocrine mechanism that differentiates (is defeminized) under the influence of gonadal steroids during early life, as observed in rats, or reflect another type of stable difference (e.g., in testicular or pituitary physiology) that could interfere with the positive and negative feedback (Baum et al., 1985; Gladue, 1985). In both cases, it appears that male homosexuality is associated with a differ-ence in functioning of the hypothalamic–pituitary–testicular axis.

10.3.4 Brain Activation in Response to Putative Pheromones

In many species of mammals, body odor of males or females plays a critical role in the control of sexual behavior by modulating the attractiveness of sexual partners and stimulating (or inhibiting in some situations) hormonal activity underlying reproduction. These odors are grouped under the term of pheromones.

The existence of pheromones in humans is controversial and it is beyond the scope of this presentation to discuss this problem (see Meredith, 2001; Halpern & Martinez-Marcos, 2003; Wysocki & Preti, 2004 for further information) but suffice to say that a few compounds present ei-

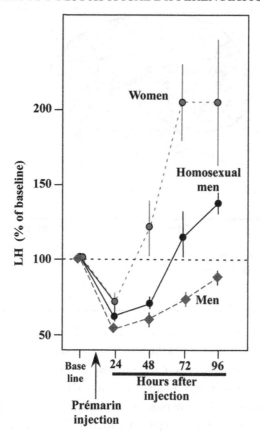

FIGURE 17: Changes in blood concentrations of luteinizing hormone (LH) in response to injection of a single dose of Premarin in heterosexual men and women and in homosexual men. Redrawn from Gladue et al. (1984).

ther in the axillary sweat or in the urine can be detected in humans, consciously or not, and influence various responses such as a choice of location (chairs in a waiting room, from a series of identical urinals) and/or modify the physiology and mood of subjects, or even their sexual motivation.

We saw in Chapter 6 that a perinatal treatment of male rat pups with an aromatase inhibitor produces adult males who are bi- or homosexual (Bakker et al., 1993) and who, contrary to control males, display an increased brain activation (measured by induction of the Fos protein) in response to cage litter soiled by male urine (Bakker et al., 1996a).

In humans, two compounds derived from steroids are considered by some (but this is disputed) as human pheromones. These compounds are the androgen 1, 4-androstadien-3-one (AND) and the estrogenic compound estra-1, 3, 5 (10),16-tetraene-3-ol (EST). AND was identified in the

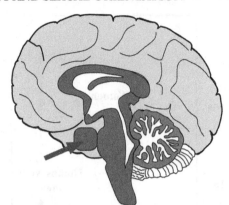

	Men		Women	
	Hetero	Homo	Hetero	Homo
AND	–	+	+	–
EST	+	–	–	–

FIGURE 18: Exposure to the putative pheromones AND and EST (see text) differentially activates the hypothalamic area of males and females and the response is modified based on their sexual orientation. The upper figure presents a schematic view of a sagittal section of human brain showing the region (arrow and red square) where metabolic activations were detected by medical imaging. The table indicates whether these hypothalamic activations are present (+) or absent (–) depending on the sex of the individual, his/her sexual orientation and the compound used. AND = compound produced by men; EST = compound produced by women. Redrawn from data in Savic et al. (2005) and Berglund et al. (2006).

axillary sweat in men, while EST is present in the urine of women. Using two separate techniques of medical imaging, positron emission tomography (PET) and functional nuclear magnetic resonance imaging (fMRI), a group of researchers from the Karolinska Institute in Stockholm asked whether these compounds produce a differential activation of the brain in men and women and if so whether this activation is affected by the sexual orientation of the subjects. In a first step, differential brain activation was identified in men and women (Savic et al., 2001). Exposure to EST, but not to the AND induced in males an activation of the anterior hypothalamus. In women hypothalamic activation was detected after exposure to AND but not to EST (Figure 18).

Interestingly, these activations were profoundly altered in homosexual subjects. The preoptic area and hypothalamus were significantly activated by exposure to AND in gay men as they are in women but not in heterosexual men. In contrast, exposure to the EST, the female stimulus that activated hypothalamic areas in heterosexual men, was not active in homosexual men (Savic et al.,

2005) (see Figure 18). Similar changes were observed in parallel in homosexual women who were no longer activated by AND as normally seen in heterosexual women (Berglund et al., 2006).

These results demonstrate that the brain of homosexual men and women is differentially activated by odors with sexual connotations as compared to the brain of heterosexual subjects of the same sex. However, these data do not indicate whether the differential activation is the cause or consequence of the sexual orientation. One can indeed imagine that the atypical AND- or EST-induced brain activation in homosexual subjects results from their atypical early hormonal imprinting and causes their attraction to subjects of the same sex. Alternatively, it is also conceivable that being gay or lesbian produces a sensitization to specific odors from individuals of the same sex and that this learning is reflected at the level of brain activity detected by *in vivo* imaging. By analogy with animal data, one might however be tempted to choose the first of these interpretations. It should also be noted that the brain activations discussed here were observed after exposure to concentrations of AND or EST that are presumably much higher than physiological concentrations encountered in the real world. More research on this topic is therefore warranted.

10.4   IN CONCLUSION

Taken together, studies described in this chapter indicate that multiple behavioral, morphological and physiological characteristics suggest an inappropriate masculinization in male homosexuals and a hyper-masculinization in lesbians. Based on available data, it is however likely that the prenatal hormonal influences do not represent a strict determinism. The changes associated with homosexuality are only statistically related to this sexual orientation and only explain a significant but limited part of the observed variance. Either other factors play an important role or the relationships between embryonic hormones, sexual orientation and the various characteristics associated with this orientation are obscured by specific problems (dose-response, critical period, . . .) that have not been identified so far. More work would be needed to discriminate between these interpretations.

These studies have however another major interest. They indicate that homosexuality is not only a change in sexual orientation but is also associated quite reproducibly to morphological, physiological or behavioral differences unrelated to sexual behavior. These associations indicate in an indirect but fairly conclusively manner that homosexuality cannot be a lifestyle choice, as many people think. It is difficult to imagine how free choices could induce a change in the relative size of the fingers or a differential functioning of the inner ear. These differences obviously precede the appearance of homosexual orientation, and the most parsimonious interpretation is that the same (hormonal?) cause is responsible for all of them.

·   ·   ·   ·   ·

CHAPTER 11

Brain Differences Associated with Sexual Orientation

Several limbic and hypothalamic nuclei involved in the control of sexual behavior are sexually differentiated. They are either larger in males than in females (e.g., sexually dimorphic nucleus [SDN] of the preoptic area, bed nucleus of the stria terminalis [BNST]) or more developed in females than in males (antero-ventral nucleus of the anterior hypothalamus [AVPV], ventromedial nucleus of the hypothalamus [VMN]) (De Vries & Simerly, 2002). It is established that the volume of several of these nuclei (SDN, BNST, AVPV) is determined in rats by the action of sex hormones during the embryonic or immediately postnatal period and cannot be modified in adulthood. If the same was true in humans, the volume of these nuclei could then represent a reliable marker of the hormonal milieu in which an individual has developed. If these nuclei were then modified according to the sexual orientation of individuals, they would provide an additional argument concerning the validity of the hormonal hypothesis of sexual orientation.

Most of the structures of interest have unfortunately a small volume and still cannot be identified by currently available imaging techniques (nuclear magnetic resonance imaging [MRI] or positron emission tomography [PET]). Only histological studies of post-mortem tissue can provide useful information. Because experimental material is obviously difficult to obtain, only a few studies have been performed. Very interesting results have nevertheless been identified concerning the suprachiasmatic nucleus (SCN), bundles of fibers connecting the two cerebral hemispheres and the sexually dimorphic nucleus of preoptic area (SDN-POA).

11.1 THE SUPRACHIASMATIC NUCLEUS

A study of 18 heterosexual and 10 homosexual men showed that a small group of neurons in the hypothalamus called the suprachiasmatic nucleus (SCN) is significantly larger (1.7 times) and contains twice as many neurons in homosexual than in heterosexual men (Swaab & Hofman, 1990). This study is mentioned because it was the first to demonstrate a neuroanatomical difference related to sexual orientation and as such suggested that homosexuality is associated with specific biological changes. However, this neuroanatomical difference provides no information about the mechanisms that lead to homosexuality. The SCN is a group of cells, mainly related to the control of circadian rhythms that is only indirectly involved in the control of certain aspects of reproduction (see Swaab & Hofman, 1990 for discussion). There are no data suggesting a direct involvement in the control of sexual orientation (Kruijver et al., 1993). In addition, there is no evidence that the SCN volume is sexually differentiated (greater or smaller in men than in women). Thus, the difference observed between homosexual and straight men does not provide any argument that might suggest a lack of masculinization of the brain.

11.2 THE ANTERIOR COMMISSURE

The anterior commissure connects the two cerebral hemispheres at the level of the anterior hypo-thalamus. A study of the brains of 90 subjects (heterosexual men and women and homosexual men) showed that this connection is significantly more developed in women than in men when measured in the midsagittal plane (Allen & Gorski, 1992). Interestingly, the size of this commissure is also larger among homosexual than in heterosexual men and even than in women (Figure 19).

This neuroanatomical difference observed in gay men is an additional argument suggesting that male homosexuality may be related to an atypical sexual differentiation. This structure has, however, no direct link with sexuality, but may contribute to some gender differences in various cognitive tasks such as visuo-spatial skills or verbal tasks and sex differences of functional lateraliza-

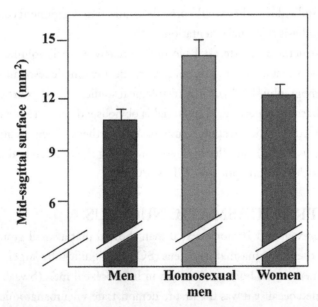

FIGURE 19: Size of the brain anterior commissure estimated from its surface in the midsagittal plane in men and women and in homosexual men. Redrawn from Allen & Gorski (1992).

tion (the brain of men seems more strongly lateralized than in females). Because gay men are more similar to women for some of these criteria (see previous chapter), the large size of their anterior commissure correlates nicely with their cognitive abilities. Note, however, that some studies failed to confirm these cognitive differences between hetero- and homosexual men, and they must be treated with caution.

11.3   THE CORPUS CALLOSUM

In 1982, biologist Christine de Lacoste-Utamsing and her colleague Ralph Holloway published an article based on the post-mortem study of human brains showing that the corpus callosum, another connection between the left and right cerebral hemispheres, is sexually differentiated (DeLacoste-Utamsing & Holloway, 1982). In particular, a part of the corpus callosum called the splenium was shown to be larger in women than in men. This work received enormous attention, and since the structure involved is large and can now be measured by brain imaging in living subjects several studies have re-analyzed this question. Some could not reproduce the sex difference while others confirmed it. Its existence thus remains controversial but the detail of this debate is peripheral to the topic of this book.

In contrast, the more recently identified relationship between the shape of the corpus callosum and sexual orientation deserves our full attention. Male homosexuality is associated with a significant change in the manual lateralization: gay men are more often not right-handed (left-handed or ambidextrous) than heterosexuals. Researchers have questioned whether this unusual lateralization is associated with a difference in the size and shape of the corpus callosum. This magnetic resonance imaging study of the corpus callosum was performed in twelve homosexual and ten heterosexual men who were all strictly right-handed in order to analyze effects of sexual orientation without the confounding factor of manual lateralization. They observed that the isthmus, a subregion of the posterior corpus callosum, is significantly larger in homosexuals suggesting that they have, like women, a brain asymmetry that is less pronounced than heterosexual men. Moreover, in this study, statistical analysis correctly identified in 21 out of 22 cases the sexual orientation of the subject based only on the size of the isthmus of the corpus callosum but also on the basis of different cognitive tests already mentioned in the previous chapter (Witelson et al., 2007). It is also important to remember that manual handedness develops very early, usually before birth (Hepper et al., 1991). This feature cannot be a personal choice. The fact that handedness is correlated with sexual orientation thus suggests that sexual orientation is also determined early in life.

11.4   THE SDN-POA

As explained in Chapter 6, partner preference in male rats can be modified experimentally and irreversibly by altering the hormonal milieu just before or immediately after birth (Bakker et al.,

1993; Bakker et al., 1996b). This change from a female to a male partner preference is associated with a decrease in the volume of the sexually dimorphic nucleus of the preoptic area (Houtsmuller et al., 1994).

    One or more sexually dimorphic nuclei are also present in the preoptic area of the human brain. A preoptic nucleus larger in men than in women was first identified by Dick Swaab and colleagues at the Netherlands Institute for Brain Research in Amsterdam (Swaab & Fliers, 1985). This difference is not present at birth but gradually develops during the first 10–15 years of life and persists until an advanced age (> 80 years) although its amplitude tends to decrease with age (Swaab & Hofman, 1988). Analysis of the size of this nucleus in the brains of homosexual men, however,

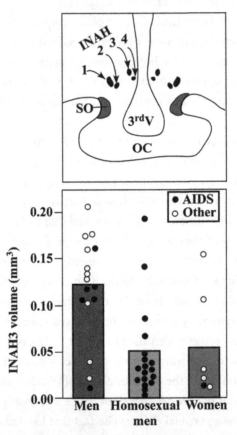

FIGURE 20: Schematic representation of the human hypothalamus illustrating the position of the 4 nuclei of the interstitial Anterior Hypothalamus (INAH) in relation to the 3rd ventricle (3rdV), the chiasma of the optic nerves (OC) and the supra-optic nucleus (SO). The lower figure represents the volumes of INAH3 measured in men, in gay men and in women who died of AIDS or of another cause. The bars represent the average for each group of subjects. Redrawn from data in LeVay (1991).

identified no difference in volume compared to heterosexual men, although this study highlighted a difference in the volume of another nucleus as a function of sexual orientation of the subjects (see above section on the supra-chiasmatic nucleus).

Researchers from the laboratory of Roger Gorski (who had identified the SDN-POA of rats) at the University of California, Los Angeles also analyzed the human preoptic area and described four distinct cell condensations ("nuclei") in this region. They called these nuclei the INAH 1 to 4 for the Interstitial Nuclei of the Anterior Hypothalamus. Two of these nuclei were larger in men than in women (INAH 2 and 3) (Allen et al., 1989), and it was shown a little later that the INAH3 is significantly smaller in homosexual men than heterosexual men. Its average size is in fact similar to the size of INAH3 observed among women (LeVay, 1991) (Figure 20).

An independent study performed later on a different set of brains confirmed the numerically reduced size of INAH3 in homosexual men compared to heterosexual subjects but the magnitude of the difference was lower than in the first study of LeVay and did not reach statistical significance (Byne et al., 2001). This study also identified a strong trend ($p =.057$) for a greater cell density (more cells per unit volume) in the INAH3 of homosexual than in heterosexual men. The homosexual nucleus had the same number of neurons as in heterosexual men, but these neurons were closer to each other on average (possibly because they formed fewer synapses during development).

11.4.1 Interpretation

The presence of a smaller INAH3 in male homosexuals acquires a specially important meaning when connecting this observation to the animal studies that have shown that:

1. Lesions of the SDN-POA induce a change in male partner preference, from strictly heterosexual to either homosexual or bi-sexual (Paredes & Baum, 1995; Paredes et al., 1998).
2. The larger size of SDN-POA in males compared to females is determined exclusively by the action of sex steroids during embryonic and postnatal life (masculinization by testosterone and its estrogenic metabolite, estradiol) (Jacobson et al., 1981; Arnold & Gorski, 1984; Rhees et al., 1990). Hormone treatments in adulthood have no effect on the size of the nucleus.
3. Perinatal treatment with an aromatase inhibitor that modifies the sexual orientation of male rats (Bakker et al., 1993; Bakker et al., 1996b) in parallel reduces the size of the SDN of the preoptic area (Houtsmuller et al., 1994).
4. In rams, a smaller SDN of the preoptic area is associated with a homosexual partner preference (Roselli et al., 2004a; Roselli et al., 2004b). The size of this nucleus in sheep is also controlled by the action of testosterone during embryonic life (Roselli et al., 2007).

If the human INAH3 is homologous to and plays the same role as the rat SDN-POA or the ovine SDN, its smaller size in homosexual men would support the theory of an early hormonal origin of homosexuality. The small INAH3 of homosexuals would be, at the very minimum, the signature of an atypical embryonic hormonal environment and, given the role of the SDN-POA in the control of partner preference, could even be considered as one causal determinant of homosexuality. Several considerations should however temper the tendency to extrapolate too quickly from animals to men.

1. The structure of the human POA is complicated. Four different INAH (1–4) have been identified. Three of them were found to be larger in men than in women: the SDN of Swaab DF & Fliers (Swaab & Fliers, 1985) is probably homologous to INAH1 in Allen et al. (1989) who failed to confirm the sex difference in the volume of this nucleus (presumably to due to differences in the ages of the subjects included in the two studies, (Garcia-Falgueras & Swaab, 2008), and INAH-2 and -3 (Allen et al., 1989). INAH-3 only changes as a function of sexual orientation (LeVay, 1991), but its relation to the rat or ovine SDN is not clear. Although INAH-3 is located roughly in the same place as the SDN of rats, no manipulation (e.g., lesion) can be performed to confirm that INAH3 indeed controls sexual orientation in humans.

2. It is impossible to confirm that the human INAH-3 volume is determined exclusively by embryonic hormones as demonstrated for the SDN in rats and probably in sheep. A study demonstrated that the size of INAH3 is slightly, though not significantly, reduced in human males castrated in adulthood for medical reasons (prostate cancer). INAH3 volume may thus reflect, at least in part, hormonal status in adulthood (Garcia-Falgueras & Swaab, 2008).

3. It was argued that the small size of the INAH3 in homosexual men was the result of the fact they all died from AIDS. Authors of these studies were, however, aware of this limitation. The LeVay study (LeVay, 1991) included heterosexuals who also died from AIDS but they had an INAH-3 size larger than the average INAH-3 of homosexuals. In the study of Byne and colleagues (Byne et al., 2001), there was also no influence of AIDS on the INAH-3 volume in heterosexual men (9 who died from AIDS and 22 from other causes).

4. Even if we can dismiss all these objections, the studies of LeVay and Byne cannot fully answer the fundamental question of whether the smaller INAH-3 in homosexual men is the cause or the consequence of their sexual orientation. We spontaneously think that changes in structure or function of the brain determine changes in behavior but the reverse causal link is also possible. Brain structure is indeed plastic and can be affected by past experiences of a subject. For example, London taxi drivers have a larger hippocampus, a

brain area involved in spatial orientation, than control subjects and motor areas controlling the fingers of the left hand are larger in violinists than in matched subjects not playing this instrument.

The smaller size of the INAH-3 in homosexuals could thus theoretically be induced by an aspect of their behavior or lifestyle, rather than be the cause of their orientation. This argument can not be formally rejected, but it must be noted that all cases of brain plasticity induced by behavior in the nervous system that have been described both in animals and in humans concern the cerebral hemispheres but never the hypothalamus and limbic system. Rather, the hypothalamus appears to be controlled by intrinsic physiological mechanisms based largely on hormonal changes. It is not excluded that in the future, researchers will discover changes in the hypothalamus induced by experience, but in the current state of knowledge, no data can support this hypothesis. Together, these data associated with the animal studies reviewed before support the idea that the small INAH-3 precedes rather than follows the appearance of homosexuality, which supports the hormonal theory of homosexuality.

Two interpretations of this neuroanatomical correlate of sexual orientation are thus possible and likely. The small size of the homosexual INAH-3 might simply represent a signature of the early hormonal environment in which these individuals developed. The size of the INAH3 and sexual orientation would have been induced both by the presence of abnormal embryonic hormonal conditions (probably a too low concentration of testosterone), but these two features would not be directly related in a causal manner. Alternatively, considering that lesions of the preoptic area (including SDN) of rats or ferrets change their sexual partner preference, these two features might be causally related: the low level of early testosterone action would have caused the development of a small INAH3 that would in turn be (one of) the cause(s) of the homosexual orientation.

Even if this last scenario turned out to be true, it must be emphasized that a decreased volume of INAH-3 cannot be the only cause of homosexuality. There is indeed quite a bit of overlap between the volumes observed in homosexual and heterosexual men even if the average volumes differ statistically. The size of this nucleus thus cannot be the sole cause of homosexuality. Changes in INAH-3 may predispose to a given sexual orientation but not be the only causal factor.

· · · ·

CHAPTER 12

The Origin of Endocrine Differences Between Embryos

It must be realized that if one accepts the endocrine model of sexual orientation (Chapter 7) according to which male homosexuality would relate to a lower than usual action of testosterone during early life and female homosexuality to an increased action of this steroid (or its metabolites), one is still left with the question of why testosterone action was changed in some subjects as compared to others during their early life. Three groups of mechanisms can be considered and have been investigated to various degrees in an attempt to answer this question. Testosterone secretion or action could be affected by external events that were experienced by the mothers during their pregnancy. Alternatively, genetic differences between subjects could be the direct cause of their endocrine difference that leads to the development of a homosexual orientation. Finally, mothers may mount an immune response toward their male embryo and alter their brain development either directly or through changes in their endocrine physiology. Let us consider these possibilities in sequence.

12.1 EFFECTS OF EXTERNAL EVENTS

The focus of study here has been on the potential role of maternal stress. Ingeborg and Byron Ward and their colleagues from Villanova University in Pennsylvania showed that stressing pregnant rats by immobilization in a highly illuminated area affects the sexual differentiation of young males that will be born from these mothers (Ward, 1972; 1984). These males have at birth an anogenital distance (distance between the base of the penis and anus that is normally smaller in females than in males) that is lower than normal and in adulthood, they mount the females less frequently and are even capable of presenting female-typical behaviors such as lordosis in response to sexual advances of other males (Ward & Ward, 1985). The volume of their SDN-POA is correlatively smaller than in control males and thus closer to female-typical values (Anderson et al., 1985; Kerchner & Ward, 1992). In summary, they show a partial lack of masculinization and defeminization. Analysis of the endocrine status of these stressed embryos revealed that their blood level of testosterone was reduced and, in addition, that aromatase activity (conversion of testosterone into estradiol) was inhibited in their preoptic area (Weisz, 1983; Ward, 1984; Jimbo et al., 1998). These hormonal changes induced by stress and the associated increase in circulating levels of corticosterone were clearly responsible for their incomplete sexual differentiation (see Chapter 5).

In a retrospective analysis of men born in Berlin during World War II, Günter Dörner, a researcher working at the time in East Germany, showed the existence of a significant peak in frequency of gay men among cohorts of boys born in Berlin between 1942 and 1946 (Dörner, 1980; Dörner et al., 1980; Dörner et al., 1983). This increase did not appear to be linked to an increased reporting or detection of homosexuals linked to the appearance around 1970 (when children were 20–30 years old) of a more tolerant attitude towards homosexuality. Indeed, this peak is transient and returns to baseline around 1948–1949. It is thus specific to pregnancies having taken place during the second world war. It is easy to imagine that many of these mothers had experienced extremely stressful events during pregnancy. A detailed analysis of the questionnaires indicated a relationship between highly stressful life events that mothers remembered and the likelihood that their boy born in this period be homosexual (Dörner et al., 1983). These data therefore suggest that stress experienced during pregnancy may increase the likelihood of homosexual orientation in males probably due to an interference with the production or action of androgens as described in the rat.

Note however that a) the animal studies of Ward and colleagues showed an effect of stress on the type of behavior (male- or female-specific) performed and size of the SDN-POA but not sexual orientation, b) a few more recent studies failed to reproduce the correlation between maternal stress and homosexual behavior observed by Dörner (Schmidt & Clement, 1990; Bailey et al., 1991) or produced only equivocal data (Ellis et al., 1988). The importance of stress experienced by mothers in these more recent studies is, however, probably much lower than in the Dörner study investigating World War II but in the absence of replication, it must be admitted that it remains unclear whether prenatal stress is able to affect sexual orientation in humans.

It is also important to emphasize here that a huge amount of data has accumulated during the last few decades indicating that a variety of chemical compounds released in the environment affect the endocrine physiology and in particular the process of sexual differentiation in many animal species and possibly also in humans (Panzica et al., 2007; Li et al., 2008; Gore & Patisaul, 2010). They are grouped under the term endocrine disruptors. There is to our knowledge no study that has tried to relate this exposure to environmental endocrine disruptors to changes in sexual orientation in humans, but this is a definite possibility that would deserve to be investigated (Li et al., 2008).

12.2   GENETIC DIFFERENCES

Genetic differences could affect the synthesis of steroid hormones or their activity in the brain of the embryo through changes for example in receptor expression or in the synthesis and activity of steroid metabolizing enzymes. This possibility has been tested to a certain extent but no positive information has been collected in this search. It is obvious that genes coding for the androgen or estrogen receptors or for aromatase cannot be completely lacking or non-functional in homosexual subjects because these subjects would then suffer from multiple physiological (e.g., lack of ovulation in females) and morphological deficits (males without androgen receptor would look like women). However, smaller variations (mutation or polymorphism) in these genes might slightly alter the function of the corresponding receptors or enzymes and in this way modify brain differentiation specifically without affecting peripheral morphology and physiology. Several studies investigated this possibility in relation to the androgen receptor and to aromatase but to date none of them has been able to identify variations in these genes that are significantly linked to sexual orientation (Macke et al., 1993; Kruijver et al., 2001; DuPree et al., 2004). Various arguments nevertheless suggest a significant genetic contribution to sexual orientation. Numerous epidemiological studies have indeed demonstrated a concordance of sexual orientation directly correlated with genetic relatedness. For example, if, in a given population, a son is gay, between 20% and 25% of his brothers will share the same orientation, as compared to 4–6% in a control population (Diamond, 1993; Rahman & Wilson, 2003a). Similarly, lesbians have a greater probability than heterosexual women of having a homosexual sister.

Twins studies showed that this correspondence in sexual orientation probably does not reflect a communality of postnatal experiences (psycho-social factors) but rather genetic similarity. The agreement of sexual orientation is indeed markedly and very significantly higher in monozygotic twins (identical twins; over 60% concordance in orientation according to a review by Diamond, 1993) than in di-zygotic twins (fraternal twins born to different ova and sperm; around 15%; see Figure 21). Another study in the 1990s identified similar rates of concordance (monozygotic twins: 52%, dizygotic twins: 22% and adoptive brothers: 11%; see Bailey & Pillard, 1991).

Two recent studies (Bailey et al., 2000; Kendler et al., 2000) based on a large populations of twins and a more rigorous selection of subjects reported however a lower concordance between monozyotic twins (between 20% and 30%) and a concordance between sexual orientation of twins was also detected in females (0.58, heritability (Kirk et al., 2000); see Rahman & Wilson, 2003a for more details). Whatever the precise value hiding behind these estimates, all studies show a better concordance in identical than in fraternal twins indicating that in social conditions typical of Western societies, 30% to 60% of the variance in sexual orientation in humans should have a genetic origin (LeVay & Hamer, 1994; Rahman & Wilson, 2003a; Swaab, 2007).

Although this genetic contribution has been identified many years ago, the responsible gene(s) remain unknown at present. During this search for relevant genes, it was shown that sexual orientation in men tends to be transmitted through matriarchal lineage: a gay man has a higher probability

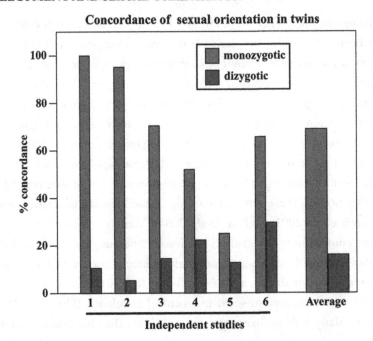

FIGURE 21: Percentage of concordance of homosexual orientation among true (monozygotic) or false (dizygotic) twins observed in 6 independent studies involving a total of 270 pairs of true and 271 pairs of false fraternal twins and average of the results. Note that a concordance of 100% does not indicate the presence of 100% of homosexuals in the population, but only that if in a twin pair one of the subjects was homosexual, the other was as well. Drawn based on data in Swaab (2007).

of having gay men among his ancestors on the maternal side (uncles, cousins), but not the paternal side. This was originally interpreted as a sign of inheritance through gene(s) located on the X chromosome (the only one inherited in a systematically different way from the father and mother) and one study indeed identified a linkage with markers located in the sub-telomeric region of the long arm of the X chromosome, a region called Xq28 (Hamer et al., 1993). This association with Xq28 was replicated in two subsequent studies (Hu et al., 1995; Sanders & Dawood, 2003) but not in a fourth one which had however a lower statistical power (Rice et al., 1999) (see also (Bocklandt & Vilain, 2007)). More recent work has however suggested that the differential heritage through the matriarchal lineage could also be the result of epigenetic modifications of the expression of genes (gene inactivation) located on several other chromosomes (see Bocklandt et al., 2006; Bocklandt & Vilain, 2007; Ngun et al., 2011a). Interestingly, an epigenetic regulation of genes controlling sexual orientation may also explain why there is no complete concordance between orientation of monozygotic (true) twins who have in theory all of their DNA in common (Fraga et al., 2005). This hypothesis opens up a new large field of potential investigations.

In women, studies have also identified an increased rate of "non-heterosexuality" (sum of homo- and bisexuality) in nieces and cousins of the paternal lineage of lesbians. This transmission is also consistent with a link to the X chromosome, but it could come from the father as well as the mother. Other interpretations are also possible and the interpretation of these data remains difficult (see Rahman & Wilson, 2003a).

To make a long story short, the existence of a genetic contribution to the determination of sexual orientation is now firmly established, but the specific gene(s) that are implicated in this process have not been identified so far. Whether this partial genetic control is mediated by alterations of steroid action during ontogeny or more directly by a sexually differentiated expression of specific genes has not been determined.

12.3   THE OLDER BROTHER'S EFFECT: AN IMMUNE REACTION?

For over 20 years, Ray Blanchard, Anthony Bogaert and their colleagues have shown that in humans, there is a highly reproducible correlation between the number of older brothers of a given male subject and the probability that he is homosexual. This effect has been called the "older brothers effect." An analysis of 14 independent studies based on more than 10,000 subjects found that for each additional older brother the probability of being gay increases by 33% (Figure 22). This does not of course mean that 33% of boys who have a brother born before them are gay but that the likelihood of developing this orientation is 33% higher than in the general population for each older brother (Blanchard, 1997; 2004).

Such a correlation could have many causes and interpretations. Given the robustness of the phenomenon and because it has been reproduced in many studies on a large number of subjects, it was possible by statistical analyses to exclude most of these alternative interpretations. It has in this process been shown that the effect is not observed for younger brothers born after the probant nor for sisters born before or after the probant (older or younger, see Figure 22). The effect also does not depend on the number of brothers who were raised at the same time as the probant (social effect that would be derived from an infancy spent with a lot of other boys; Bogaert, 2006) nor on the age of the mother or father and it is not found in girls. It is still observed if subjects are raised in different "reconstituted" families (after a divorce) but not for half-brothers (born to a different mother) or adopted brothers (Blanchard et al., 2006). These data therefore exclude, as effectively as possible in humans, the influence of post-natal educational effects of numerous siblings.

The effect thus seems to relate exclusively to the sequence of births, and the only remaining explanation at present is based on a postulated immune response of the pregnant mother directed toward the male embryo. The mother would progressively develop over successive pregnancies an immune response against male embryos that secrete foreign proteins in greater quantities than

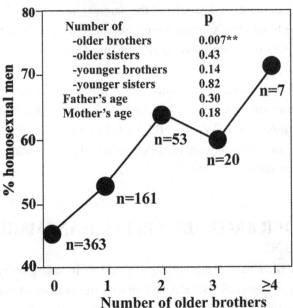

FIGURE 22: Relationship between the number of older brothers born to the same mother and the percentage of gay men in a population of 302 gay men and 302 heterosexual matched subjects used as controls. The homosexual and heterosexual subjects being matched for this study, there was therefore 50% of homosexuals in the population studied which is of course much higher than what is found in a normal population. The figure also lists the probability associated with different predictors of homosexuality in the logistic regression and the number of subjects in each category. Redrawn from Blanchard & Bogaert (1996).

female embryos (Blanchard & Bogaert, 1996; Blanchard & Klassen, 1997). This accumulation of antibodies would be similar to that seen if a Rhesus (Rh) negative mother sequentially carries several Rh positive embryos. The first embryo is almost unaffected but the mother progressively mounts an immune response and following embryos may be severely attacked by the maternal antibodies. In the case of the older brother's effect, these antibodies would affect aspects of brain development that are involved in determining sexual orientation.

Consistent with this hypothesis, the birth weight of boys with older brothers is significantly lower than the weight of matched boys with the same number of older sisters (Blanchard, 2001). Moreover, within this population of subjects with older brothers, gay men had at birth a lower body weight than heterosexual men (about 170 g of difference, which is low but significant in the sample).

The specific mechanism underlying this phenomenon has not yet been identified, but candidate proteins have been suggested as potential target(s) for this immune reaction (Ngun et al., 2011b). These candidate proteins must in theory fulfill several criteria such as 1) the fetal protein must enter the maternal circulation, 2) it must be a male-specific substance that causes an immune response in females, 3) it should be expressed in the brain and 4) it should play a role in the differentiation of the brain. According to a recent review, four proteins currently meet these criteria and are the focus of ongoing research, but it will obviously be a huge challenge to relate the expression of one or several of these proteins to the incidence of homosexuality in men and then, if this quest is successful, to identify their mechanism of action (see Ngun et al., 2011b for additional discussion).

As discussed in the previous section on genetic effects, it is conceivable that this postulated immune reaction modulates the secretion or action of steroid hormones (mainly testosterone) during embryonic life. Such a mechanism would then be compatible and actually would contribute to explain the endocrine theory of homosexuality that has been presented before. However, there is to this date no indication that this immune response is mediated by a change in endocrine physiology and a more direct, non-hormonal, action on brain development is also very likely. Nevertheless, the effect of older brothers is without any doubt an argument of great weight supporting the idea that sexual orientation is determined for a good part before birth and, in most cases, not a free choice of the adult subject.

· · · ·

CHAPTER 13

Is There a Role for Postnatal Experiential Factors?

There is a body of converging evidence suggesting a role of prenatal biological factors (hormones, genes, immune reactions) that affect adult sexual orientation. It is also clear, however, that the currently identified prenatal factors do not fully explain the incidence of homosexual orientation in men and women. This raises the question of whether additional currently unidentified prenatal factors are implicated or alternatively postnatal social factors interact with them to determine sexual orientation. Although evidence for such interactions remains scant at this time, some data are nevertheless available.

We reviewed in Chapter 6 a number of studies in rodents demonstrating that perinatal endocrine manipulations affect in a prominent and irreversible manner the sex partner preference of the treated subjects. One set of studies in birds, specifically in zebra finches (*Taeniopygia guttata*) a small songbird of the estrildid family, has confirmed this finding originally established in rodents (Adkins-Regan, 2011). It was shown that unlike control females, female zebra finches injected with estradiol will prefer to pair with other females when adults (Mansukhani et al., 1996). Very importantly however, this reversed preference is only observed if young females are raised as juveniles and young adults in all-female unisex groups. Birds similarly treated with estradiol but raised in mixed groups of males and females pair like control females with other males (Figure 23).

Surprisingly, however, treatment of young females with Fadrozole, an aromatase inhibitor, produced a similar sex reversal of partner preference (preferential pairing with females) (Adkins-Regan & Wade, 2001). Taken together, these experiments leave open a number of questions (see Adkins-Regan, 2011 for discussion) but they clearly illustrate the notion that effects of the early endocrine environment interact with effects of postnatal social experience to determine adult partner preference. This idea is also clearly supported by a broader review of the literature indicating that "sexual experience allows animals to form instrumental and Pavlovian associations that predict sexual outcome and thereby directs the strength of sexual responding" (Pfaus et al., 2003). To what extent this experience can also affect the choice of the sex partner is however poorly understood

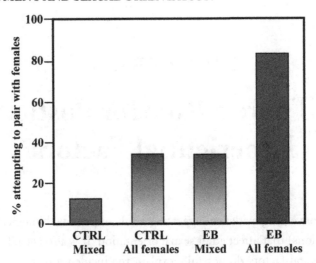

FIGURE 23: Estradiol benzoate (EB) injections in female zebra finch nestlings reverse sex partner preference so that when adults, the majority of them will attempt to pair with other females. This effect is observed only if these females lived in all female unisexual groups. Effects of EB are not observed in females raised in mixed-sex groups and rearing in these conditions also has no effect per se. Redrawn from Mansukhani et al. (1996).

and more experimental studies on a variety of animal models should be performed to document this type of interaction.

In humans, some indirect evidence has also been collected indicating that interactions with the social environment during childhood correlate with and thus potentially affect adult sexual orientation. This is suggested by the comparison of two cases of little boys who had their penis accidentally damaged and were thus reassigned as girl at the age of 7 or 18 months.

The first of these cases was actually the center of attention of the scientific but also the general press and has been at the origin of a major shift in the interpretation of the mechanisms controlling the development of gender identity and sexual orientation. This young boy known as the John/Joan case (actually called Bruce) was born in Canada in the early 1960s and was reassigned and raised as a girl after his penis was accidentally destroyed during circumcision. It was indeed thought at the time that if a child were raised in an unambiguous way as a girl, he would develop into a normal heterosexual woman in adulthood (Money & Ehrhardt, 1972). At the age of 2 years, the sex change was completed by a medical intervention. The child was surgically castrated and a rudimentary vagina was constructed from the skin of the scrotum.

In the years that followed, it seemed that the sex change was successful and that Bruce, now called Brenda, was developing as a normal girl, apart from the fact that she/he was slightly masculine in terms of playing activity (a feature of male gender role). This case thus supported the contention that culture has a decisive influence on sexual identity and gender role and denied the importance of genetics and biology on human behavior. Several decades later, biologist and sexologist Milton Diamond (University of Hawaii) and his colleague Keith Sigmundson discovered what had really happened during late childhood and puberty (Diamond & Sigmundson, 1997). The child was in fact never socialized as a normal girl and he rebelled early. He was refusing to wear feminine clothing, urinated standing up, and felt permanently that there was something wrong.

A "female puberty" was induced by treatment with estrogen, but Brenda/Bruce hated her/his breasts and refused the hormone. At the age of fourteen, he/she insisted on knowing the truth, which the parents reluctantly revealed. He/she was relieved to finally understand the root of the conflicting feelings he/she experienced. At the age of 15 years, Brenda/Bruce asked people to call him/her David and he went through additional surgery to recover a masculine identity. He underwent a double mastectomy (surgical removal of breasts) and a phalloplasty (surgical reconstruction of a penis). He was also treated with testosterone. He had always been exclusively attracted to women (sexual orientation in agreement with genetic and hormonal sex during embryonic development, not with sex of education during childhood) and, in adulthood, he married a woman with whom he had male-typical sex with his prosthesis. His story is told in a book written by John Colapinto (Colapinto, 2000).

This case provides evidence that human beings are not psycho-socially neutral at birth. In keeping with their legacy as mammals, they have trends and predispositions to react according to a male or female pattern. Sexual identity and orientation are probably influenced by social factors but also by the embryonic hormonal environment. David Reimer had obviously been exposed *in utero* to androgen levels typical of a young boy (his penis was normally formed at birth). In this case, the hormonal factors would have outweighed the social influences in determining identity and gender role.

A similar case of penile ablation has been reported but leads to a slightly different conclusion (Bradley et al., 1998). This genetic male (XY) subject accidentally lost his penis at the age of 2 months and was raised as female from the age of 7 months on. In adulthood, contrary to John/Joan, he adopted a female sex identity despite the fact that during childhood he enjoyed stereotypically masculine toys and games (partially masculinized sex role). His sexual orientation was essentially bisexual. He was however predominantly but not exclusively attracted to women. The role of education as compared to embryonic hormones was apparently more prominent in this case than in the patient studied by John Money, possibly because the sex of rearing (as female) was adopted earlier. The difference between these two cases might thus relate to the age at sex reassignment (7 vs. 18 months) and suggests that postnatal socialization might also be important in determining sexual orientation and gender identity. These are, however, only single cases, and it is therefore difficult to base general conclusions on such anecdotes.

More work would thus be needed on the possible interaction between prenatal endocrine factors and the influence of postnatal socialization, but this endeavor is particularly complicated given the relatively limited numbers of subjects that were exposed in a documented manner to atypical endocrine prenatal environment. Two laboratories are currently engaged in studies on this topic focusing on the best-documented cases of atypical endocrine environment during prenatal life, i.e., the girls affected by congenital adrenal hyperplasia (CAH, see Chapter 9). These girls have provided compelling evidence that prenatal androgenization leads to a masculinization of cognition, infant play (toy preferences and playmate selection) and various personality characteristics. Furthermore, recent work has shown that testosterone concentrations measured in maternal blood during pregnancy or in amniotic fluid during gestation predicts male-typical behavior in physiologically normal boys (Hines et al., 2002; Auyeung et al., 2009). In addition, there is evidence that gender atypical behavior during childhood is a strong predictor of homosexuality and also possibly of trans-sexuality (Berenbaum & Hines, 1992; Bailey, 2003). Finally, non-heterosexual behavior (homo- or bisexuality) is significantly increased in CAH girls (Chapter 9). Taken together, these data strongly suggest that prenatal androgens play a substantial role in the control of various aspects of human behavior including sexual orientation.

Other factors including socialization and interaction with the peers may however play a role too and the magnitude of these postnatal influences probably varies across behavioral characteristics. The development of sex-typical traits in humans involves social and cognitive mechanisms related to gender identification. Specifically, girls and boys learn early on that they are male or female and develop by social imitation behaviors that are associated with their gender. Recent evidence from the laboratory of Dr. Melissa Hines at the University of Cambridge (UK) indicates that social interactions are implicated in the development of the specific interest for objects or toys that are preferentially associated with their own sex. Most interestingly, girls with CAH do not appear to respond as typically developing (unaffected) children do to models of the same sex and to labels as to which objects are for children of their own sex (Hines, 2012). Prenatal testosterone could thus influence directly brain mechanisms controlling sexually differentiated behavioral characteristics including sexual orientation but could in addition affect the development of these differentiated traits by modulating the response to postnatal socializing factors. In this context, it should also be reminded (see Chapter 9) that a substantial proportion of CAH women experience discomfort with their own sex and genitalia and the increased incidence of homosexuality in this group could thus relate to the difficulties associated with penetrative heterosexual interactions. This would represent another indirect way through which prenatal testosterone could influence sexual orientation (masculinized genitalia → difficult heterosexual relationships → shift towards homosexuality).

Recent work from the laboratory of Sheri Berenbaum at Pennsylvania State University also supports the idea that the postnatal social environment could affect the outcome of CAH by showing that the degree of prenatal androgenization in CAH girls interacts with the postnatal social pressure to behave as men or women (boy or girl) in determining cognitive traits and sex-typical interests in children (Beltz et al., 2012).

Research on this interaction between pre- and postnatal factors in the control of sexual orientation is however still in its infancy, and more experimental data would certainly be needed before any firm conclusion can be presented.

. . . .

CHAPTER 14

Conclusions

The research summarized in this presentation clearly provides a bundle of converging elements that strongly suggest, even if they do not prove, that sexual orientation in human is (strongly) influenced by prenatal factors including hormones, genes and possibly immune reactions of the mother toward her male child. To briefly summarize the arguments, it is now relatively firmly established that in a variety of animal species:

a) the preoptic area plays a key role in the control of partner preference and partner preference is largely determined by hormonal exposure during early life (pre- or immediately postnatal period),

b) in parallel, embryonic sex steroids differentiate the size of several brain structures, including the sexual dimorphic nucleus of the POA and the sex-typical size of this nucleus is not affected by steroid hormones in adulthood,

c) a case of spontaneous strict homosexual partner preference has been described in sheep and in this species the ram-orientated males have a sexually dimorphic nucleus of the preoptic area that has the same size as in females. The size of this nucleus, which is correlated with partner preference, is determined by the action of testosterone during embryonic life.

Clinical and epidemiologic studies strongly suggest that these mechanisms identified in animals are still present in humans and play a critical role in the determination of sexual orientation. It is indeed clear that the sexual differentiation of human genital structures during embryonic life is controlled by testosterone (and its metabolites) and that the action of the steroid is not limited to the periphery but also markedly affects the development of brain and behavior. Several diseases that affect the functioning of the endocrine system during fetal life are associated with an increased incidence of homosexuality in men or women. These pathologies clearly suggest a significant role of embryonic hormones in determining sexual orientation.

In addition, it has been noted that homosexuality does not simply represent a different sexual orientation. Homosexual subjects display a number of complex differences when compared to heterosexual subjects. In homosexual men, these features are shifted in a female direction, whereas in homosexual women, they are shifted in a male direction. It could be suggested that in some cases, these differences are the consequence of homosexuality. This could be the case for some of the behavioral differences (e.g., in verbal fluency) or even in brain structure (the smaller INAH3 of gay men could be a consequence of their life style even if animal research strongly suggests a control by embryonic steroids). However, these features also include morphological (ratio of length of fingers 2D:4D, length of long bones) or physiological characteristics (oto-acoustic emissions produced by the inner ear), and it is difficult in these cases to imagine how these characteristics could be influenced by or be a consequence of sexual orientation.

These differences provide strong support to the idea that sexual orientation is not a matter of free choice performed during adolescence or in adulthood (how could a homosexual subject change the length of its fingers or its inner ear physiology?). Furthermore, since the homologous characteristics are in a variety of animal species influenced by and differentiate sexually following prenatal exposure to testosterone, they support the notion that homosexual subjects were during their embryonic life exposed to atypical concentrations of this steroid.

Of particular importance in this context is the difference concerning the size of the sexually dimorphic nucleus of preoptic area called INAH3. This nucleus is larger in heterosexual men than in women and has in homosexual males an average women size. The mechanisms that control the development of this nucleus in humans are unknown, but its volume does not seem to significantly depend on the hormonal status in adulthood. In rats and probably in sheep, the size of this nucleus is determined irreversibly by the action of embryonic sex steroids, and its lesion in adults changes sexual orientation. The difference associated with masculine homosexuality in humans is thus highly suggestive again of an atypical early exposure to testosterone.

Together, these data support the notion that prenatal sex steroids influence or even determine sexual orientation in humans as they do in rodents. There are, however, a number of problems and limitations to this conclusion.

14.1 LIMITATIONS OF THE PRENATAL HORMONAL THEORY OF HOMOSEXUALITY

On the one hand, all described embryonic endocrine diseases, including those that strongly modify the secretion or action of sex steroids, statistically increase the incidence of homosexuality but do not modify sexual orientation of all individuals. For example, congenital adrenal hyperplasia increases the proportion of girls who become lesbians in adulthood from 1–10% in a control population to 35–40% in some studies, but this still leaves more than half of the population unaffected (with a heterosexual orientation).

On the other hand, all morphological or physiological correlates suggesting that homosexuality is associated with abnormal sexual differentiation provide only an imperfect vision of hormonal events that should have happened during the embryonic period. Even if the average values of the correlated features are statistically modified, there is a huge degree of individual variance resulting in an overlap between values measured in homosexual and heterosexual subjects. For example, the INAH3 of homosexual men is on average statistically smaller than in heterosexual men but with the overlap observed, some homosexual subjects have a larger nucleus than the heterosexual men that have the smallest nuclei. Furthermore, these characteristics are not changed in all studies and are not always observed in both sexes (2D:4D ratio or oto-acoustic emissions changed in lesbians but not gay men), which raises the question of why these correlations with independent markers of homosexuality are not stronger and more reproducible.

Several factors could potentially contribute to obscure the relationships between homosexuality and these markers.

1. Human homosexuality is a complex phenomenon that could have different causes in different individuals. Within a given sex, homosexuals have very different phenotypes. Lesbians for examples can be very masculine (butch) or feminine (femme) whereas gay men can show very masculine or in contrast some effeminate features. It is possible that prenatal hormones determine homosexuality in some (types of) subjects but not in others so that the morphological, physiological and behavioral traits under study would be changed only in some homosexual individuals.

2. The classification of subjects as homo- or heterosexual (not to mention all intermediates) is often based on questionnaires or spontaneous statements and could be biased for a variety of reasons. Some people will still be reluctant to publicly acknowledge their orientation. The selection of subjects for these studies is also a critical factor and could induce a lack of reproducibility in the results. Erroneous inclusion of subjects in a given category could then dilute the relationship with the "markers" of homosexuality. Subjects recruited in places preferentially frequented by homosexuals (e.g., "gay" bars, gay pride, . . .) or as part of a survey in a University or College sample may represent different types of homosexuality and thus potentially different causes for this sexual orientation.

3. Studies conducted so far have not determined whether the hormonal difference proposed to lead to homosexuality is the embryonic concentration of circulating testosterone or the brain sensitivity to the action of this steroid. If the level of circulating testosterone is affected during a critical period of development, orientation might be affected as well as all other androgen-dependent responses that are developing at the same time. If it is in contrast, the sensitivity of the brain to testosterone that is changed while circulating concentrations are normal, one could expect to find adult homosexual individuals in whom most physiological or morphological characteristics are identical to those of heterosexuals.

Only neural or behavioral characteristics that depend on the same mechanism of sensitivity to androgens as sexual orientation will be affected. Note that these two options are not mutually exclusive, so that correlations with peripheral changes would be present in some homosexuals (if circulating testosterone was affected) but not in other subjects (if brain sensitivity testosterone is changed).

4. Dose–response relationships that link the concentration of circulating testosterone during embryonic development with the mechanism(s) controlling sexual orientation and its various correlates are not known and could be different. If human sexual orientation is more sensitive to small variations in testosterone levels than other hormonal responses described above (to which it is sometimes correlated), homosexual individuals could have other aspects of their phenotype affected (strong hormonal changes) or not (more subtle hormonal changes).

5. Similarly, the critical period when hormones potentially affect sexual orientation and its various correlates might be partially different as observed for various responses studied in animals. Hormonal perturbations limited in time could thus affect sexual orientation without affecting the other characteristics.

These last 3 possibilities are directly derived from animal studies but cannot be easily tested in humans since obvious ethical reasons prohibit deliberate manipulations of embryonic hormones. Studies are limited to the analysis of clinical cases with spontaneous genetic or hormonal changes. These so-called invoked experiments are always difficult to interpret because variables that are not taken into account may be the cause of observed changes. The long latency (15 to 20 years) between the postulated embryonic hormonal events and their effects (homosexual or heterosexual orientation) also makes any prospective study almost impossible and retrospective studies are potentially affected by sampling biases complicating their interpretation.

It must also be noted that any prenatal endocrine phenomenon that leads to homosexuality is obviously not fully deterministic but presumably modulates the probability of occurrence of this orientation in relation to specific postnatal factors that are to this date not identified (interaction with parents, teachers or society in general). A predisposition to homosexuality induced by prenatal endocrine influences could be expressed only in specific postnatal circumstances (see Chapter 13) that have not been identified.

14.2 WHAT COULD BE THE ORIGIN OF ENDOCRINE DIFFERENCES?

It must also be noted that if this endocrine theory of sexual orientation is correct, it is only moving the potential explanation one step further. If there is a difference in steroid production or steroid action between homosexual and heterosexual subjects, the causes of these differences still need to be identified. As mentioned in the text, genetic factors controlling steroid secretion or steroid action might be at the origin of these differences. Studies using a candidate gene approach (analysis of steroid receptors, of aromatase, . . .) have, to this date, failed to identify such genetic factors.

However, various arguments, including multiple studies in twins clearly indicate a concordance of sexual orientation that can only be explained easily by a genetic control (but not an absolute determinism) of sexual orientation. It has indeed been repeatedly shown that if one twin is homosexual, the probability that the second twin is also gay is much higher in "real" twins originating from one egg and thus sharing the same genetic material than among the "false" fraternal twins.

This genetic contribution to male homosexuality (mediated or not by endocrine mechanisms) would be preferentially inherited through the maternal line and a region of the X chromosome accordingly displays variability that is significantly associated with sexual orientation. A maternal heritage could also be explained by differences in maternal gene imprinting (epigenetic regulation of gene expression) but, to date, genes whose structure or expression would be linked to homosexuality have not been isolated. This genetic contribution to sexual orientation presumably involves a large number of genes and will therefore be very difficult to identify.

Note finally that the likelihood of homosexuality increases by 33% for each older brother born from the same mother and is accompanied by a small but statistically significant decrease in weight at birth. This is the best-documented factor leading to the development of homosexuality in males, but the underlying mechanism has not been formally identified although an immune reaction of the mother toward male embryos during successive pregnancies is highly suspected. Whether this immune reaction affects sexual orientation by modifying endocrine action in the developing fetus has never been considered.

14.3 PRENATAL BIOLOGICAL FACTORS VERSUS POSTNATAL SOCIAL INFLUENCES

There are thus serious limitations and many gray areas associated with the endocrine theory of sexual orientation. These limiting factors are however well known to the authors of the studies reported here and more research on these topics would be needed to solve unanswered questions. It must be noted that the alternative explanations of homosexuality that are based on psychoanalytical, psychological or sociological analyses of post-natal social factors, if they are attractive at first glance and consequently broadly accepted, have received little, if any, experimental support. In particular, psychoanalytical interpretations are based mainly on suites of anecdotes published by Sigmund Freud that have no general validity and do not resist experimental scrutiny (Van Rillaer, 1980; Bénesteau, 2002; Dufresne, 2007; Onfray, 2010). For example, the absence of father has no effect on the occurrence of male homosexuality (no increase among children raised by a single mother) contrary to predictions that would be derived of the Oedipus complex theories broadly publicized by Freud.

Theories attributing a major role to the early sexual experiences or relationships with parents similarly find no support in controlled scientific studies and are at odds with many quantitative observations. There is also no increase in the incidence of homosexuality in societies of Malaysia, where homosexual experiences are the rule among the adolescent boys or in boys who have been raised in boarding schools where homosexual relations were present on a regular basis during adolescence (no effect of first sexual experiences).

These negative results do not necessarily imply a complete lack of influence of the postnatal environment on the development of sexual orientation. More research is definitely warranted on this topic. It is clear that even if the prenatal factors discussed here explain a significant part of the variance associated with sexual orientation and create important predispositions towards homo- or heterosexuality, sexual orientation is not expressed at birth. It will only become apparent during and after puberty and there are thus plenty of opportunities for postnatal experiences to influence this phenotypic characteristic. The social environment *per se* probably does not play a critical role in the determination of sexual orientation because if this was the case, it is likely that experimental studies would have to this date identified the critical factors. However, it is clear also that currently identified prenatal factors do not fully determine sexual orientation; they only explain a part of the observed variance and thus create at best strong predispositions. Thus, either additional control factors need to be identified (may be specific genes or specific endocrine variables) or more attention should be given to the interaction between prenatal predispositions and currently unidentified social postnatal factors. The recent suggestion that CAH girls exposed prenatally to abnormally high concentrations of androgens are differentially sensitive to social influences and social pressure (see Chapter 13) makes it very likely that such interactions may actually be of critical importance. This suggests new ways in which the prenatal hormonal environment could affect sexual orientation by changing reactions to postnatal social interactions.

. . . .

References

Adkins-Regan, E. (2011) Neuroendocrine contributions to sexual partner preference in birds. *Front Neuroendocrinol*, **32**, pp. 155–63.

Adkins-Regan, E. & Wade, J. (2001) Masculinized sexual partner preference in female zebra finches with sex-reversed gonads. *Horm Behav*, **39**, pp. 22–8.

Allen, L.S. & Gorski, R.A. (1992) Sexual orientation and the size of the anterior commissure in the human brain. *Proc Natl Acad Sci USA*, **89**, pp. 7199–202.

Allen, L.S., Hines, M., Shryne, J.E. & Gorski, R.A. (1989) Two sexually dimorphic cell groups in the human brain. *J Neurosci*, **9**, pp. 497–506.

Anderson, D.K., Rhees, R.W. & Fleming, D.E. (1985) Effects of prenatal stress on differentiation of the sexually dimorphic nucleus of the preoptic area (SDN-POA) of the rat brain. *Brain Res*, **332**, pp. 113–8.

Arendash, G.W. & Gorski, R.A. (1983) Effects of discrete lesions of the sexually dimorphic nucleus of the preoptic area or other medial preoptic regions on the sexual behavior of male rats. *Brain Res Bull*, **10**, pp. 147–54.

Arnold, A.P. & Gorski, R.A. (1984) Gonadal steroid induction of structural sex differences in the central nervous system. *Ann Rev Neurosci*, **7**, pp. 413–42.

Auyeung, B., Baron-Cohen, S., Ashwin, E., Knickmeyer, R., Taylor, K., Hackett, G. & Hines, M. (2009) Fetal testosterone predicts sexually differentiated childhood behavior in girls and in boys. *Psychol Sci*, **20**, pp. 144–8.

Bailey, J.M., Dunne, M.P. & Martin, N.G. (2000) Genetic and environmental influences on sexual orientation and its correlates in an Australian twin sample. *J Pers Social Psychol* **78**, pp. 524–36.

Bailey, J.M. & Pillard, R.C. (1991) A genetic study of male sexual orientation. *Arch Gen Psych*, **48**, pp. 1089–96.

Bailey, J.M., Willerman, L. & Parks, C. (1991) A test of the maternal stress theory of human male homosexuality. *Arch Sex Behav*, **20**, pp. 277–93.

Bailey, M.J. (2003) *The Man Who Would Be Queen: The Science of Gender-Bending and Transsexualism*. Joseph Henry Press.

Bakker, J., Baum, M.J. & Slob, A.K. (1996a) Neonatal inhibition of brain estrogen synthesis alters adult neural Fos responses to mating and pheromonal stimulation in the male rat. *Neuroscience*, **74**, pp. 251–60.

Bakker, J., De Mees, C., Douhard, Q., Balthazart, J., Gabant, P., Szpirer, J. & Szpirer, C. (2006) Alpha-fetoprotein protects the developing female mouse brain from masculinization and defeminization by estrogens. *Nat Neurosci*, **9**, pp. 220–6.

Bakker, J., Honda, S., Harada, N. & Balthazart, J. (2002a) The aromatase knock-out mouse provides new evidence that estradiol is required during development in the female for the expression of sociosexual behaviors in adulthood. *J Neurosci*, **22**, pp. 9104–12.

Bakker, J., Honda, S., Harada, N. & Balthazart, J. (2002b) Sexual partner preference requires a functional aromatase (cyp19) gene in male mice. *Horm Behav*, **42**, pp. 158–71.

Bakker, J., Van Ophemert, J. & Slob, A.K. (1993) Organization of partner preference and sexual behavior and its nocturnal rhythmicity in male rats. *Behav. Neurosci*, **107**, pp. 1049–58.

Bakker, J., Van Ophemert, J. & Slob, A.K. (1996b) Sexual differentiation of odor and partner preference in the rat. *Physiol Behav*, **60**, pp. 489–94.

Balthazart, J. (2011) *The biology of homosexuality*. Oxford University Press, New York.

Balthazart, J. & Ball, G.F. (2007) Topography in the preoptic region: differential regulation of appetitive and consummatory male sexual behaviors. *Front Neuroendocrinol*, **28**, pp. 161–78.

Bancroft, J. (1995) *The pharmacology of sexual function and dysfunction*. Elsevier Science, Amsterdam.

Baron-Cohen, S. (2004) *The Essential Difference: Men, Women and the Extreme Male Brain*. Penguin Press Science, London.

Baron-Cohen, S. (2006) *Prenatal testosterone in mind*. MIT Press, Cambridge, MA.

Baum, M.J., Carroll, R.S., Erskine, M.S. & Tobet, S.A. (1985) Neuroendocrine response to estrogen and sexual orientation. *Science*, **230**, pp. 960–1.

Becker, J.B., Berkley, K.J., Geary, N., Hampson, E., Herman, J.P. & Young, E.A. (2008) *Sex differences in the brain. From Genes to behavior*. Oxford University Press, Oxford.

Becker, J.B., Breedlove, S.M., Crews, D. & McCarthy, M.M. (2002) *Behavioral Endocrinology*. MIT Press, Cambridge MA.

Beltz, A.M., K.L., B. & Berenbaum, S.A. (2012) Dose-dependent prenatal androgen influences on gender development in girls with congenital adrenal hyperplasia. *Abst Soc Behav Neuroendocrinol (Madison WI)*, P3.45.

Bénesteau, J. (2002) *Mensonges Freudiens*. Pierre Mardaga, Wavre (Belgique).

Berenbaum, S.A., Bryk, K.K., Nowak, N., Quigley, C.A. & Moffat, S. (2009) Fingers as a marker of prenatal androgen exposure. *Endocrinology*, **150**, pp. 5119–24.

Berenbaum, S.A., Duck, S.C. & Bryk, K. (2000) Behavioral effects of prenatal versus postnatal androgen excess in children with 21-hydroxylase-deficient congenital adrenal hyperplasia. *The Journal of Clinical Endocrinology and Metabolism*, **85**, pp. 727–33.

Berenbaum, S.A. & Hines, M. (1992) Early androgens are related to childhood sex-typed toy preferences. *Psychol Sci*, **3**, pp. 203–6.

Berenbaum, S.A. & Snyder, E. (1995) Early hormonal influences on childhood sex-typed activity and playmate preferences: implications for the development of sexual orientation. *Developmental Psychology*, **31**, pp. 31–42.

Berglund, H., Lindstrom, P. & Savic, I. (2006) Brain response to putative pheromones in lesbian women. *Proc Natl Acad Sci USA*, **103**, pp. 8269–74.

Berta, P., Hawkins, J.R., Sinclair, A.H., Taylor, A., Griffiths, B.L., Goodfellow, P.N. & Fellous, M. (1990) Genetic evidence equating SRY and the testis-determining factor. *Nature*, **348**, pp. 448–50.

Beyer, C., Vidal, N. & Mijares, A. (1970) Probable role of aromatization in the Induction of estrous behavior by androgens in the ovariectomized rabbit. *Endocrinology*, **87**, pp. 1386–9.

Blanchard, R. (1997) Birth order and sibling sex ratio in homosexual versus heterosexual males and females. *Ann Rev Sex Res*, **8**, pp. 27–67.

Blanchard, R. (2001) Fraternal birth order and the maternal immune hypothesis of male homosexuality. *Horm Behav*, **40**, pp. 105–14.

Blanchard, R. (2004) Quantitative and theoretical analyses of the relation between older brothers and homosexuality in men. *Journal of Theoretical Biology*, **230**, pp. 173–87.

Blanchard, R. & Bogaert, A.F. (1996) Homosexuality in men and number of older brothers. *Am J Psychiatry*, **153**, pp. 27–31.

Blanchard, R., Cantor, J.M., Bogaert, A.F., Breedlove, S.M. & Ellis, L. (2006) Interaction of fraternal birth order and handedness in the development of male homosexuality. *Horm Behav*, **49**, pp. 405–14.

Blanchard, R. & Klassen, P. (1997) H-Y antigen and homosexuality in men. *J Theor Biol*, **185**, pp. 373–8.

Blanchard, R. & Lippa, R.A. (2007) Birth order, sibling sex ratio, handedness, and sexual orientation of male and female participants in a BBC internet research project. *Arch Sex Behav*, **36**, pp. 163–76.

Blaustein, J.D. (2012) Animals Have a Sex, and so Should Titles and Methods Sections of Articles in Endocrinology. *Endocrinology*, **153**, pp. 2539–40.

Bocklandt, S., Horvath, S., Vilain, E. & Hamer, D.H. (2006) Extreme skewing of X chromosome inactivation in mothers of homosexual men. *Hum Genet*, **118**, pp. 691–4.

Bocklandt, S. & Vilain, E. (2007) Sex differences in brain and behavior: hormones versus genes. *Adv Genet*, **59**, pp. 245–66.

Bodo, C. & Rissman, E.F. (2007) Androgen receptor is essential for sexual differentiation of responses to olfactory cues in mice. *Eur J Neurosci*, **25**, pp. 2182–90.

Bodo, C. & Rissman, E.F. (2008) The androgen receptor is selectively involved in organization of sexually dimorphic social behaviors in mice. *Endocrinology*, **149**, pp. 4142–50.

Bogaert, A.F. (2006) Biological versus nonbiological older brothers and men's sexual orientation. *Proc Natl Acad Sci U S A*, **103**, pp. 10771–4.

Bradley, S.J., Oliver, G.D., Chernick, A.B. & Zucker, K.J. (1998) Experiment of nurture: ablatio penis at 2 months, sex reassignment at 7 months, and a psychosexual follow-up in young adulthood. *Pediatrics*, **102**, pp. 1–5.

Breedlove, S.M. (2010) Minireview: Organizational hypothesis: instances of the fingerpost. *Endocrinology*, **151**, pp. 4116–22.

Brock, O. & Bakker, J. (2011) Potential contribution of prenatal estrogens to the sexual differentiation of mate preferences in mice. *Horm Behav*, **59**, pp. 83–9.

Brown, W.M., Finn, C.J. & Breedlove, S.M. (2002a) Sexual dimorphism in digit-length ratios of laboratory mice. *Anat Rec*, **267**, pp. 231–4.

Brown, W.M., Finn, C.J., Cooke, B.M. & Breedlove, S.M. (2002b) Differences in finger length ratios between self-identified "butch" and "femme" lesbians. *Arch Sex Behav*, **31**, pp. 123–7.

Brown, W.M., Hines, M., Fane, B.A. & Breedlove, S.M. (2002c) Masculinized finger length patterns in human males and females with congenital adrenal hyperplasia. *Horm Behav*, **42**, pp. 380–6.

Byne, W., Tobet, S., Mattiace, L.A., Lasco, M.S., Kemether, E., Edgar, M.A., Morgello, S., Buchsbaum, M.S. & Jones, L.B. (2001) The interstitial nuclei of the human anterior hypothalamus: An investigation of variation with sex, sexual orientation, and HIV status. *Horm Behav*, **40**, pp. 86–92.

Cahill, L. (2006) Why sex matters for neuroscience. *Nature reviews. Neuroscience*, **7**, pp. 477–84.

Campbell, B.C., Prossinger, H. & Mbzivo, M. (2005) Timing of pubertal maturation and the onset of sexual behavior among zimbabwe school boys. *Arch. Sexual Behav*, **34**, pp. 505–16.

Chivers, M.L., Rieger, G., Latty, E. & Bailey, J.M. (2004) A sex difference in the specificity of sexual arousal. *Psychological Science*, **15**, pp. 736–44.

Colapinto, J. (2000) *As nature made him: The boy that was raised as a girl*. Harper Collins, New York.

Collaer, M.L. & Hines, M. (1995) Human behavioral sex differences: a role for gonadal hormones during early development? *Psychol. Bull.*, **118**, pp. 55–107.

Davidson, J.M., Camargo, C.A. & Smith, E.R. (1979) Effects of androgen on sexual behavior in hypogonadal men. *The Journal of clinical endocrinology and metabolism*, **48**, pp. 955–8.

De Jonge, F.H., Louwerse, A.L., Ooms, M.P., Evers, P., Endert, E. & Van De Poll, N.E. (1989) Lesions of the SDN-POA inhibit sexual behavior of male Wistar rats. *Brain Res Bull*, **23**, pp. 483–92.

De Vries, G.J., Rissman, E.F., Simerly, R.B., Yang, L.Y., Scordalakes, E.M., Auger, C.J., Swain, A., Lovell-Badge, R., Burgoyne, P.S. & Arnold, A.P. (2002) A model system for study of sex chromosome effects on sexually dimorphic neural and behavioral traits. *J Neurosci*, **22**, pp. 9005–14.

De Vries, G.J. & Simerly, R.B. (2002) Anatomy, development, and function of sexually dimorphic neuroal circuits in the mammalian brain. In Pfaff, D.W., Arnold, A.P., Etgen, A.M., Fahrbach, S.E., Rubin, R.T. (eds) *Hormones, Brain and Behavior*. Academoc Press, San Diego, CA, pp. 137–91.

DeLacoste-Utamsing, C. & Holloway, R.L. (1982) Sexual dimorphism in the human corpus callosum. *Science*, **216**, pp. 1431–2.

Diamond, M. (1993) Some genetic considerations in the development of sexual orientation. In Haug, M., Whalen, R.E., Aron, C., Olsen, K.L. (eds.), *The development of sex differences and similarities in behavior*. Kluwer Academic Publishers, Dordrecht, pp. 291–309.

Diamond, M. & Sigmundson, H.K. (1997) Sex reassignment at birth: long-term review and clinical implications. *Arch Pediatrics and Adolescent Medicine*, **151**, pp. 298–304.

Dittmann, R.W., Kappes, M.E. & Kappes, M.H. (1992) Sexual behavior in adolescent and adult females with congenital adrenal hyperplasia. *Psychoneuroendocrinology*, **17**, pp. 153–70.

Dörner, G. (1969) Zur Frage einer neuroendocrinen Pathogenese, Prophylaxe und Therapie angeborenen Sexualdeviationen. *Deutsche Medizinische Wochenschrift*, **94**, pp. 390–6.

Dörner, G. (1972) *Sexualhormonabhängige Gehirndifferenzierung und Sexualität*. Springer, Berlin, Heidelberg, New York.

Dörner, G. (1976) *Hormones and Brain Differentiation*. Elsevier, Amsterdam, Oxford, New York.

Dörner, G. (1980) Sexual differentiation of the brain *Vitamins and hormones*. Academic Press, Inc., pp. 325–81.

Dörner, G., Geier, T., Ahrens, L., Krell, L., Munx, G., Sieler, H., Kittner, E. & Muller, H. (1980) Prenatal stress as possible aetiogenetic factor of homosexuality in human males. *Endokrinologie*, **75**, pp. 365–8.

Dörner, G., Schenk, B., Schmiedel, B. & Ahrens, L. (1983) Stressful events in prenatal life of bi- and homosexual men. *Exp Clin Endocrinol*, **81**, pp. 83–7.

Dörner, G. & Staudt, J. (1969) Structural changes in the hypothalamic ventromedial nucleus of the male rat, following neonatal castration and androgen treatment. *Neuroendocrinol*, **4**, pp. 278–81.

Dufresne, T. (2007) *Against Freud. Critics talk back.* Stanford University Press, Stanford, CA.

DuPree, M.G., Mustanski, B.S., Bocklandt, S., Nievergelt, C. & Hamer, D.H. (2004) A candidate gene study of CYP19 (aromatase) and male sexual orientation. *Behav Genet*, **34**, pp. 243–50.

Ehrhardt, A.A., Meyer-Bahlburg, H.F., Rosen, L.R., Feldman, J.F., Veridiano, N.P., Zimmerman, I. & McEwen, B.S. (1985) Sexual orientation after prenatal exposure to exogenous estrogen. *Arch Sex Behav*, **14**, pp. 57–77.

Ellis, L., Ames, M.A., Peckham, W. & Ahrens, L. (1988) Sexual orientation of human offspring may be altered by severe maternal stress during pregnancy. *J Sex Res*, **25**, pp. 152–7.

Ellis, L., Hershberger, S., Field, E., Wersinger, S., Pelis, S., Geary, D., Palmer, C., Hoyenga, K., Hetsroni, A. & Karadi, K. (2008) *Sex differences: summarizing more than a century of scientific research.* Psychology Press, New York.

Ellis, L., Hoffman, H. & Burke, D.M. (1990) Sex, sexual orientation and criminal and violent behavior. *Personality and Individual Differences*, **11**, pp. 1207–11.

Etgen, A.M. & Pfaff, D.W. (2009) *Molecular mechanisms of hormone actions on behavior.* Elsevier, Amsterdam.

Fine, C. (2011) *Delusions of Gender. The real science behind sex differences.* Icon Books, London.

Fraga, M.F., Ballestar, E., Paz, M.F., Ropero, S., Setien, F., Ballestar, M.L., Heine-Suner, D., Cigudosa, J.C., Urioste, M., Benitez, J., Boix-Chornet, M., Sanchez-Aguilera, A., Ling, C., Carlsson, E., Poulsen, P., Vaag, A., Stephan, Z., Spector, T.D., Wu, Y.Z., Plass, C. & Esteller, M. (2005) Epigenetic differences arise during the lifetime of monozygotic twins. *Proc Natl Acad Sci USA*, **102**, pp. 10604–9.

Freud, S. (1905/1975) *Three essays on the theory of sexuality.*

Freund, K.W. (1974) Male homosexuality: an analysis of the pattern. In Lorraine, J.A. (ed.), *Understanding homosexuality: its biological and psychological bases.* Elsevier, Amsterdam.

Garcia-Falgueras, A. & Swaab, D.F. (2008) A sex difference in the hypothalamic uncinate nucleus: relationship to gender identity. *Brain*, **131**, pp. 3132–46.

Gladue, B.A. (1985) Neuroendocrine response to estrogen and sexual orientation. *Science*, **230**, p. 961.

Gladue, B.A. & Bailey, J.M. (1995) Aggressiveness, competitiveness, and human sexual orientation. *Psychoneuroendocrinol*, **20**, p. 475.

Gladue, B.A., Beatty, W.W., Larson, J. & Staton, R.D. (1990) Sexual orientation and spatial ability in men and women. *Psychobiol*, **18**, pp. 101–8.

Gladue, B.A., Green, R. & Hellman, R.E. (1984) Neuroendocrine response to estrogen and sexual orientation. *Science*, **225**, pp. 1496–9.

Gobrogge, K.L., Breedlove, S.M. & Klump, K.L. (2008) Genetic and environmental influences on 2D:4D finger length ratios: a study of monozygotic and dizygotic male and female twins. *Arch Sex Behav*, **37**, pp. 112–8.

Gore, A.C. & Patisaul, H.B. (2010) Neuroendocrine disruption: historical roots, current progress, questions for the future. *Front Neuroendocrinol*, **31**, pp. 395–9.

Gorski, R.A. (1984) Critical role of the medial preoptic area in the sexual differentiation of the brain. In De Vries, G.J., De Bruin, J.P.C., Uylings, H.B.M., Corner, M.A. (eds.), *Sex differences in the brain*. Elsevier, Amsterdam, pp. 129–46.

Gorski, R.A., Gordon, J.H., Shryne, J.E. & Southam, A.M. (1978) Evidence for a morphological sex difference within the medial preoptic area of the rat brain. *Brain Res.*, **148**, pp. 333–46.

Goy, R.W. & McEwen, B.S. (1980) *Sexual differentiation of the brain*. The MIT Press, Cambridge, MA.

Hajjar, R.R., Kaiser, F.E. & Morley, J.E. (1997) Outcomes of long-term testosterone replacement in older hypogonadal males: a retrospective analysis. *The Journal of clinical endocrinology and metabolism*, **82**, pp. 3793–6.

Hall, L.S. & Kimura, D. (1995) Sexual orientation and performance on sexually dimorphic motor tasks. *Arch Sex Behav*, **24**, pp. 395–407.

Hall, L.S. & Love, C.T. (2003) Finger-length ratios in female monozygotic twins discordant for sexual orientation. *Arch Sex Behav*, **32**, pp. 23–8.

Halpern, C.T., Udry, J.R. & Suchindran, C. (1998) Monthly measures of salivary testosterone predict sexual activity in adolescent males. *Archives of Sexual Behavior*, **27**, pp. 445–65.

Halpern, C.T., Udry, R. & Suchindran, C. (1997) Testosterone Predicts Initiation of Coitus in Adolescent Females. *Psychosomatic Medicine*, **59**, pp. 161–71.

Halpern, M. & Martinez-Marcos, A. (2003) Structure and function of the vomeronasal system: an update. *Prog Neurobiol*, **70**, pp. 245–318.

Hamer, D.H., Hu, S., Magnuson, V.L., Hu, N. & Pattatucci, A.M.L. (1993) A linkage between DNA markers on the X chromosome and male sexual orientation. *Science*, **261**, pp. 321–7.

Henley, C.L., Nunez, A.A. & Clemens, L.G. (2009) Estrogen treatment during development alters adult partner preference and reproductive behavior in female laboratory rats. *Horm Behav*, **55**, pp. 68–75.

Hepper, P.G., Shahidullah, S. & White, R. (1991) Handedness in the human fetus. *Neuropsychologia*, **29**, pp. 1107–11.

Hines, M. (2003) Sex steroids and human behavior: prenatal androgen exposure and sex-typical play behavior in children. *Ann NY Acad Sci*, **1007**, pp. 272–82.

Hines, M. (2004) *Brain gender*. Oxford University Press, Oxford.

Hines, M. (2006) Prenatal testosterone and gender-related behaviour. *Eur J Endocrinol*, **155 Suppl 1**, pp. S115–21.

Hines, M. (2010) Sex-related variation in human behavior and the brain. *Trends in cognitive sciences*, **14**, pp. 448–56.

Hines, M. (2011) Prenatal endocrine influences on sexual orientation and on sexually differentiated childhood behavior. *Front Neuroendocrinol*, **32**, pp. 170–82.

Hines, M. (2012) Gender development and Behavior. *6th Int. Symp. On the Biology of Vertebrate Sex Determination (Kona, Hawaii)*, pp. 1–2.

Hines, M., Brook, C. & Conway, G.S. (2004) Androgen and psychosexual development: core gender identity, sexual orientation and recalled childhood gender role behavior in women and men with congenital adrenal hyperplasia (CAH). *J Sex Res*, **41**, pp. 75–81.

Hines, M., Fane, B.A., Pasterski, V.L., Mathews, G.A., Conway, G.S. & Brook, C. (2003) Spatial abilities following prenatal androgen abnormality: targeting and mental rotations performance in individuals with congenital adrenal hyperplasia. *Psychoneuroendocrinology*, **28**, pp. 1010–26.

Hines, M., Johnston, K.J., Golombok, S., Rust, J., Stevens, M. & Golding, J. (2002) Prenatal stress and gender role behavior in girls and boys: a longitudinal, population study. *Horm Behav*, **42**, pp. 126–34.

Hines, M. & Kaufman, F.R. (1994) Androgen and the development of human sex-typical behavior: rough-and-tumble play and sex of preferred playmates in children with congenital adrenal hyperplasia (CAH). *Child Dev*, **65**, pp. 1042–53.

Houtsmuller, E.J., Brand, T., de Jonge, F.H., Joosten, R.N., van de Poll, N.E. & Slob, A.K. (1994) SDN-POA volume, sexual behavior, and partner preference of male rats affected by perinatal treatment with ATD. *Physiol Behav*, **56**, pp. 535–41.

Hu, S., Pattatucci, A.M., Patterson, C., Li, L., Fulker, D.W., Cherny, S.S., Kruglyak, L. & Hamer, D.H. (1995) Linkage between sexual orientation and chromosome Xq28 in males but not in females. *Nat Genet*, **11**, pp. 248–56.

Iijima, M., Arisaka, O., Minamoto, F. & Arai, Y. (2001) Sex differences in children's free drawings: a study on girls with congenital adrenal hyperplasia. *Horm Behav*, **40**, pp. 99–104.

Imperato-McGinley, J. (1994) 5 alpha-reductase deficiency: human and animal models. *Eur Urol*, **25 Suppl 1**, pp. 20–3.

Imperato-McGinley, J., Miller, M., Wilson, J.D., Peterson, R.E., Shackleton, C. & Gajdusek, D.C. (1991) A cluster of male pseudohermaphrodites with 5 alpha-reductase deficiency in Papua New Guinea. *Clin Endocrinol (Oxf)*, **34**, pp. 293–8.

Imperato-McGinley, J. & Zhu, Y.S. (2002) Androgens and male physiology the syndrome of 5alpha-reductase-2 deficiency. *Mol Cell Endocrinol*, **198**, pp. 51–9.

Jacobson, C.D., Csernus, V.J., Shryne, J.E. & Gorski, R.A. (1981) The influence of gonadectomy, androgen exposure, or a gonadal graft in the neonatal rat on the volume of the sexually dimorphic nucleus of the preoptic area. *J Neurosci*, **1**, pp. 1142–7.

Jimbo, M., Okubo, K., Toma, Y., Shimizu, Y., Saito, H. & Yanaihara, T. (1998) Inhibitory effects of catecholamines and maternal stress on aromatase activity in the fetal rat brain. *J Obstet Gynaecol Res*, **24**, pp. 291–7.

Jordan-Young, R.M. (2010) *Brain storm. The flaws in the science of sex differences*. Harvard University Press, Cambridge MA.

Jost, A. (1985) Organogenesis and endocrine cytodifferentiation of the testis. *Arch Anat Microsc Morphol Exp*, **74**, pp. 101–2.

Jost, A., Vigier, B., Prepin, J. & Perchellet, J.P. (1973) Studies on sex differentiation in mammals. *Recent progress in hormone research*, **29**, pp. 1–41.

Kelley, D.B. & Pfaff, D.W. (1978) Generalizations from comparative studies on neuroanatomical and endocrine mechanisms of sexual behaviour. In Hutchison, J.B. (ed) *Biological determinants of sexual behaviour*. John Wiley & Sons, Chichester, pp. 225–54.

Kendler, K.S., Thornton, L.M., Gilman, S.E. & Kessler, R.C. (2000) Sexual orientation in a U.S. national sample of twin and nontwin sibling pairs. *The American journal of psychiatry*, **157**, pp. 1843–6.

Kerchner, M. & Ward, I.L. (1992) SDN-MPOA volume in male rats is decreased by prenatal stress, but is not related to ejaculatory behavior. *Brain Res*, **581**, pp. 244–51.

Kinsey, A.C., Pomeroy, W.R. & Martin, C.E. (1948) *Sexual behavior in the human male*. W.B. Saunders Company, Philadelphia.

Kinsey, A.C., Pomeroy, W.R., Martin, C.E. & Gebhard, P.H. (1953) *Sexual behavior in the human female*. Saunders, Philadelphia.

Kirk, K.M., Bailey, J.M., Dunne, M.P. & Martin, N.G. (2000) Measurement models for sexual orientation in a community twin sample. *Behav Genet*, **30**, pp. 345–56.

Kraemer, B., Noll, T., Delsignore, A., Milos, G., Schnyder, U. & Hepp, U. (2006) Finger length ratio (2D:4D) and dimensions of sexual orientation. *Neuropsychobiology*, **53**, pp. 210–4.

Kruijver, F.P.M., Balesar, R., Espila, A.M., Unmehopa, U.A. & Swaab, D.F. (2002) Estrogen receptor-alpha distribution in the human hypothalamus in relation to sex and endocrine status. *J Comp Neurol*, **454**, pp. 115–39.

Kruijver, F.P.M., Balesar, R., Espila, A.M., Unmehopa, U.A. & Swaab, D.F. (2003) Estrogen-receptor-beta distribution in the human hypothalamus: Similarities and differences with ERalpha distribution. *J Comp Neurol*, **466**, pp. 251–77.

Kruijver, F.P.M., de Jonge, F.H., van den Broek, W.T., van der Woude, T., Endert, E. & Swaab, D.F. (1993) Lesions of the suprachiasmatic nucleus do not disturb sexual orientation of the adult male rat. *Brain Research*, **624**, pp. 342–46.

Kruijver, F.P.M., Fernández-Guasti, A., Fodor, M., Kraan, E.M. & Swaab, D.F. (2001) Sex differences in androgen receptors of the human mamillary bodies are related to endocrine status rather than to sexual orientation or transsexuality. *The Journal of clinical endocrinology and metabolism*, **86**, pp. 818–27.

Lalumiere, M.L., Blanchard, R. & Zucker, K.J. (2000) Sexual orientation and handedness in men and women: a meta-analysis. *Psychol Bull*, **126**, pp. 575–92.

LeVay, S. (1991) A difference in hypothalamic structure between heterosexual and homosexual men. *Science*, **253**, pp. 1034–7.

LeVay, S. (2010) *Gay, straight, and the reason why. The science of sexual orientation.* Oxford University Press, New York.

LeVay, S. & Hamer, D.H. (1994) Evidence for a biological influence in male homosexuality. *Sci Am*, **270**, pp. 44–9.

LeVay, S. & Valente, S.M. (2006) *Human Sexuallity*. Sinauer Associates Inc, Sunderland, MA.

Li, A.A., Baum, M.J., McIntosh, L.J., Day, M., Liu, F. & Gray, L.E., Jr. (2008) Building a scientific framework for studying hormonal effects on behavior and on the development of the sexually dimorphic nervous system. *Neurotoxicology*, **29**, pp. 504–19.

Lippa, R.A. (2003a) Are 2D:4D finger-length ratios related to sexual orientation? Yes for men, no for women. *Journal of personality and social psychology*, **85**, pp. 179–88.

Lippa, R.A. (2003b) Handedness, sexual orientation, and gender-related personality traits in men and women. *Arch Sex Behav*, **32**, pp. 103–14.

Lutchmaya, S., Baron-Cohen, S., Raggatt, P., Knickmeyer, R. & Manning, J.T. (2004) 2nd to 4th digit ratios, fetal testosterone and estradiol. *Early Hum Dev*, **77**, pp. 23–8.

Macke, J.P., Hu, N., Hu, S., Bailey, M., King, V.L., Brown, T., Hamer, D. & Nathans, J. (1993) Sequence variation in the androgen receptor gene is not a common determinant of male sexual orientation. *Am J Hum Genet*, **53**, pp. 844–52.

MacLaughlin, D.T. & Donahoe, P.K. (2004) Sex determination and differentiation. *N Engl J Med*, **350**, pp. 367–78.

Manning, J.T., Fink, B., Neave, N. & Szwed, A. (2006) The second to fourth digit ratio and asymmetry. *Ann Hum Biol*, **33**, pp. 480–92.

Mansukhani, V., Adkins-Regan, E. & Yang, S. (1996) Sexual partner preference in female zebra finches: The role of early hormones and social environment. *Horm Behav*, **30**, pp. 506–13.

Martin, J.T. & Nguyen, D.H. (2004) Anthropometric analysis of homosexuals and heterosexuals: implications for early hormone exposure. *Horm Behav*, **45**, pp. 31–9.

McCarthy, M.M. (2011) *Sex and the developing brain*. Morgan & Claypool Life Sciences.

McCarthy, M.M., Arnold, A.P., Ball, G.F., Blaustein, J.D. & De Vries, G.J. (2012) Sex differences in the brain: the not so inconvenient truth. *J Neurosci*, **32**, pp. 2241–7.

McCarthy, M.M. & Ball, G.F. (2008) The Neuroendocrine Control of Sex-Specific Behavior in Vertebrates: Lessons from Mammals and Birds. *Current Topics in Developmental Biology*, **83**, pp. 213–48.

McCormick, C.M. & Witelson, S.F. (1991) A cognitive profile of homosexual men compared to heterosexual men and women. *Psychoneuroendocrinology*, **16**, pp. 459–73.

McFadden, D. (2002) Masculinization effects in the auditory system. *Arch Sex Behav*, **31**, pp. 99–111.

McFadden, D. (2008) What do sex, twins, spotted hyenas, ADHD, and sexual orientation have in common? *Perspectives Psychol Sci*, **3**, pp. 309–23.

McFadden, D. & Champlin, C.A. (2000) Comparison of auditory evoked potentials in heterosexual, homosexual, and bisexual males and females. *J Assoc Res Otolaryngol*, **1**, pp. 89–99.

McFadden, D. & Pasanen, E.G. (1998) Comparison of the auditory systems of heterosexuals and homosexuals: click-evoked otoacoustic emissions. *Proc Natl Acad Sci USA*, **95**, pp. 2709–13.

McFadden, D. & Pasanen, E.G. (1999) Spontaneous otoacoustic emissions in heterosexuals, homosexuals, and bisexuals. *J Acoust Soc Am*, **105**, pp. 2403–13.

McFadden, D., Pasanen, E.G., Valero, M.D., Roberts, E.K. & Lee, T.M. (2009) Effect of prenatal androgens on click-evoked otoacoustic emissions in male and female sheep (Ovis aries). *Horm Behav*, **55**, pp. 98–105.

McFadden, D. & Shubel, E. (2002) Relative lengths of fingers and toes in human males and females. *Horm Behav*, **42**, pp. 492–500.

Meredith, M. (2001) Human vomeronasal organ function: a critical review of best and worst cases. *Chem Senses*, **26**, pp. 433–45.

Meyer-Bahlburg, H.F. (1984) Psychoendocrine research on sexual orientation. Current status and future options. *Prog Brain Res*, **61**, pp. 375–98.

Meyer-Bahlburg, H.F. (2005) Gender identity outcome in female-raised 46,XY persons with penile agenesis, cloacal exstrophy of the bladder, or penile ablation. *Arch Sex Behav*, **34**, pp. 423–38.

Meyer-Bahlburg, H.F. (2008) Male Gender Identity in an XX Individual with Congenital Adrenal Hyperplasia. *J Sex Med*.

Meyer-Bahlburg, H.F., Dolezal, C., Baker, S.W. & New, M.I. (2008) Sexual orientation in women with classical or non-classical congenital adrenal hyperplasia as a function of degree of prenatal androgen excess. *Arch Sex Behav*, **37**, pp. 85–99.

Meyer-Bahlburg, H.F., Ehrhardt, A.A. & Rosen, L.R. (1995) Prenatal estrogens and the development of homosexual orientation. *Dev Psychol*, **s31**, pp. 12–21.

Miller, G., Tybur, J. & Jordan, B.D. (2007) Ovulatory cycle effects on tip earnings by lap dancers: economic evidence for human estrus? *Evolution and Human Behavior*, **28**, pp. 375–81.

Money, J. & Ehrhardt, A.A. (1972) *Man & Woman, Boy & Girl*. Johns Hopkins University Press, Baltimore.

Money, J., Schwartz, M. & Lewis, V.G. (1984) Adult erotosexual status and fetal hormonal masculinization and demasculinization: 46,XX congenital virilizing adrenal hyperplasia and 46,XY androgen-insensitivity syndrome compared. *Psychoneuroendocrinology*, **9**, pp. 405–14.

Morrell, J.I. & Pfaff, D.W. (1978) A neuroendocrine approach to brain function: localization of sex steroid concentrating cells in vertebrate brains. *Amer Zool*, **18**, pp. 447–60.

Mosher , W.D., Chandra, A. & Jones , J. (2005) Sexual behavior and selected health measures: men and women 15–44 years of age, United States 2002. http://www.cdc/gov/nchs/data/ad/ad362.pdf.

Naftolin, F., Ryan, K.J., Davies, I.J., Reddy, V.V., Flores, F., Petro, Z., Kuhn, M., White, R.J., Takaoka, Y. & Wolin, L. (1975) The formation of estrogens by central neuroendocrine tissues. *Recent progress in hormone research*, **31**, pp. 295–319.

Naftolin, F., Ryan, K.J. & Petro, Z. (1971) Aromatization of androstenedione by the diencephalon. *The Journal of clinical endocrinology and metabolism*, **33**, pp. 368–70.

Neave, N., Menaged, M. & Weightman, D.R. (1999) Sex differences in cognition: the role of testosterone and sexual orientation. *Brain Cogn*, **41**, pp. 245–62.

Nelson, R.J. (2011) *An introduction to behavioral endocrinology*. Sinauer Associates, Sunderland, MA.

Ngun, T.C., Ghahramani, N., Sanchez, F.J., Bocklandt, S. & Vilain, E. (2011a) The genetics of sex differences in brain and behavior. *Front Neuroendocrinol*, **32**, pp. 227–46.

Ngun, T.C., Ghahramani, N., Sanchez, F.J., Bocklandt, S. & Vilain, E. (2011b) The genetics of sex differences in brain and behavior. *Front Neuroendocrinol*, 32. In press.

Nottebohm, F. & Arnold, A.P. (1976) Sexual dimorphism in vocal control areas of the songbird brain. *Science*, **194**, pp. 211–3.

Onfray, M. (2010) *Le crépuscule d'une idole. L'affabulation freudienne*. Bernard Grasset, Paris.

Panzica, G.C., Viglietti-Panzica, C., Calcagni, M., Anselmetti, G.C., Schumacher, M. & Balthazart, J. (1987) Sexual differentiation and hormonal control of the sexually dimorphic preoptic medial nucleus in quail. *Brain Res*, **416**, pp. 59–68.

Panzica, G.C., Viglietti-Panzica, C., Mura, E., Quinn, M.J., Jr., Lavoie, E., Palanza, P. & Ottinger, M.A. (2007) Effects of xenoestrogens on the differentiation of behaviorally-relevant neural circuits. *Front Neuroendocrinol*, **28**, pp. 179–200.

Paredes, R.G. & Baum, M.J. (1995) Altered sexual partner preference in male ferrets given excitotoxic lesions of the preoptic area anterior hypothalamus. *J Neurosci*, **15**, pp. 6619–30.

Paredes, R.G., Tzschentke, T. & Nakach, N. (1998) Lesions of the medial preoptic area anterior hypothalamus (MPOA/AH) modify partner preference in male rats. *Brain Res*, **813**, pp. 1–8.

Perkins, A. & Roselli, C.E. (2007) The ram as a model for behavioral neuroendocrinology. *Horm Behav*, **52**, pp. 70–7.

Pfaff, D., Arnold, A.P., Etgen, A.M., Fahrbach, S.E. & Rubin, R.T. (2002) *Hormones, brain and behavior*. Academic Press, Amsterdam.

Pfaus, J.G., Kippin, T.E. & Coria-Avila, G. (2003) What can animal models tell us about human sexual response? *Annu Rev Sex Res*, **14**, pp. 1–63.

Phoenix, C.H., Goy, R.W., Gerall, A.A. & Young, W.C. (1959) Organizational action of prenatally administered testosterone propionate on the tissues mediating behavior in the female guinea pig. *Endocrinology*, **65**, pp. 369–82.

Rahman, Q., Kumari, V. & Wilson, G.D. (2003a) Sexual orientation-related differences in prepulse inhibition of the human startle response. *Behav Neurosci*, **117**, pp. 1096–102.

Rahman, Q. & Wilson, G.D. (2003a) Born gay? The psychobiology of human sexual orientation. *Personality and Individual Differences*, **34**, pp. 1337–82.

Rahman, Q. & Wilson, G.D. (2003b) Large sexual-orientation-related differences in performance on mental rotation and judgment of line orientation tasks. *Neuropsychology*, **17**, pp. 25–31.

Rahman, Q. & Wilson, G.D. (2003c) Sexual orientation and the 2nd to 4th finger length ratio: evidence for organising effects of sex hormones or developmental instability? *Psychoneuroendocrinology*, **28**, pp. 288–303.

Rahman, Q., Wilson, G.D. & Abrahams, S. (2003b) Sexual orientation related differences in spatial memory. *J Int Neuropsychol Soc*, **9**, pp. 376–83.

Raisman, G. & Field, P.M. (1971) Sexual dimorphism in the preoptic area of the rat. *Science*, **173**, pp. 731–3.

Reiner, W.G. & Gearhart, J.P. (2004) Discordant sexual identity in some genetic males with cloacal exstrophy assigned to female sex at birth. *N Engl J Med*, **350**, pp. 333–41.

Rhees, R.W., Shryne, J.E. & Gorski, R.A. (1990) Termination of the hormone-sensitive period for differentiation of the sexually dimorphic nucleus of the preoptic area in male and female rats. *Dev Brain Res*, **52**, pp. 17–23.

Rice, G., Anderson, C., Risch, N. & Ebers, G. (1999) Male homosexuality: absence of linkage to microsatellite markers at Xq28. *Science*, **284**, pp. 665–7.

Rieger, G., Chivers, M.L. & Bailey, J.M. (2005) Sexual arousal patterns of bisexual men. *Psychological Science*, **16**, pp. 579–84.

Romano, M., Rubolini, D., Martinelli, R., Bonisoli Alquati, A. & Saino, N. (2005) Experimental manipulation of yolk testosterone affects digit length ratios in the ring-necked pheasant (Phasianus colchicus). *Horm Behav*, **48**, pp. 342–6.

Roselli, C.E., Larkin, K., Resko, J.A., Stellflug, J.N. & Stormshak, F. (2004a) The volume of a sexually dimorphic nucleus in the ovine medial preoptic area/anterior hypothalamus varies with sexual partner preference. *Endocrinology*, **145**, pp. 478–83.

Roselli, C.E., Larkin, K., Schrunk, J.M. & Stormshak, F. (2004b) Sexual partner preference, hypothalamic morphology and aromatase in rams. *Physiol Behav*, **83**, pp. 233–45.

Roselli, C.E., Reddy, R. & Kaufman, K. (2011) The development of male-oriented behavior in rams. *Front Neuroendocrinol*.

Roselli, C.E., Stadelman, H., Reeve, R., Bishop, C.V. & Stormshak, F. (2007) The ovine sexually dimorphic nucleus of the medial preoptic area is organized prenatally by testosterone. *Endocrinology*, **148**, pp. 4450–7.

Rosler, A. & Witztum, E. (1998) Treatment of men with paraphilia with a long-acting analogue of gonadotropin-releasing hormone. *N Engl J Med*, **338**, pp. 416–22.

Rubin, R.T., Reinisch, J.M. & Haskett, R.F. (1981) Postnatal gonadal steroid effects on human behavior. *Science*, **211**, pp. 1318–24.

Sanders, A.R. & Dawood, K. (2003) *Nature Encyclopedia of Life Sciences*. Nature Publishing Group, London.

Savic, I., Berglund, H., Gulyas, B. & Roland, P. (2001) Smelling of odorous sex hormone-like compounds causes sex-differentiated hypothalamic activations in humans. *Neuron*, **31**, pp. 661–8.

Savic, I., Berglund, H. & Lindstrom, P. (2005) Brain response to putative pheromones in homosexual men. *Proc Natl Acad Sci USA*, **102**, pp. 7356–61.

Schmidt, G. & Clement, U. (1990) Does peace prevent homosexuality? *Arch Sex Behav*, **19**, pp. 183–7.

Sherwin, B.B. & Gelfand, M.M. (1987) The role of androgen in the maintenance of sexual functioning in oophorectomized women. *Psychosom Med*, **49**, pp. 397–409.

Sinclair, A.H., Berta, P., Palmer, M.S., Hawkins, J.R., Griffiths, B.L., Smith, M.J., Foster, J.W., Frischauf, A.M., Lovell-Badge, R. & Goodfellow, P.N. (1990) A gene from the human sex-determining region encodes a protein with homology to a conserved DNA-binding motif. *Nature*, **346**, pp. 240–4.

Snyder, P.J., Peachey, H., Berlin, J.A., Hannoush, P., Haddad, G., Dlewati, A., Santanna, J., Loh, L., Lenrow, D.A., Holmes, J.H., Kapoor, S.C., Atkinson, L.E. & Strom, B.L. (2000) Effects of testosterone replacement in hypogonadal men. *J Clin Endocrinol Metab*, **85**, pp. 2670–7.

Swaab, D.F. (2007) Sexual differentiation of the brain and behavior. *Best Pract Res Clin Endocrinol Metab*, **21**, pp. 431–44.

Swaab, D.F. & Fliers, E. (1985) A sexually dimorphic nucleus in the human brain. *Science*, **228**, pp. 1112–5.

Swaab, D.F. & Hofman, M.A. (1988) Sexual differentiation of the human hypothalamus: Ontogeny of the sexually dimorphic nucleus of the preoptic area. *Dev Brain Res*, **44**, pp. 314–8.

Swaab, D.F. & Hofman, M.A. (1990) An enlarged suprachiasmatic nucleus in homosexual men. *Brain Res*, **537**, pp. 141–8.

Titus-Ernstoff, L., Perez, K., Hatch, E.E., Troisi, R., Palmer, J.R., Hartge, P., Hyer, M., Kaufman, R., Adam, E., Strohsnitter, W., Noller, K., Pickett, K.E. & Hoover, R. (2003) Psychosexual characteristics of men and women exposed prenatally to diethylstilbestrol. *Epidemiology*, **14**, pp. 155–60.

Tobet, S.A. & Fox, T.O. (1992) Sex differences in neuronal morphology influenced hormonally throughout life Handbook of behavioral neurobiology. Vol 11. Sexual differentiation. In Gerall, A.A., Moltz, H., Ward, I.L. (eds.). Plenum Press, New York, pp. 41–83.

Turkenburg, J.L., Swaab, D.F., Endert, E., Louwerse, A.L. & Van De Poll, N.E. (1988) Effects of lesions of the sexually dimorphic nucleus on sexual behavior of testosterone-treated female wistar rats. *Brain Res Bull*, **21**, pp. 215–24.

Tutle, G.E. & Pillard, R.C. (1991) Sexual orientation and cognitive abilities. *Arch Sex Behav*, **20**, pp. 307–18.

Udry, J.R., Billy, J.O., Morris, N.M., Groff, T.R. & Raj, M.H. (1985) Serum androgenic hormones motivate sexual behavior in adolescent boys. *Fertil Steril*, **43**, pp. 90–4.

Van Rillaer, J. (1980) *Les illusions de la psychanalyse*. Pierre Mardaga, Sprimont (Belgique).

Vidal, C. (2007) *Hommes, femmes avons-nous le même cerveau?* Le Pommier, Paris.

Wang, C., Swerdloff, R.S., Iranmanesh, A., Dobs, A., Snyder, P.J., Cunningham, G., Matsumoto, A.M., Weber, T., Berman, N. & Grp, T.G.S. (2000) Transdermal testosterone gel improves sexual function, mood, muscle strength, and body composition parameters in hypogonadal men. *J Clin Endocrinol Metab*, **85**, pp. 2839–53.

Ward, I.L. (1972) Prenatal stress feminizes and demasculinizes the behavior of males. *Science*, **175**, pp. 82–4.

Ward, I.L. (1984) The prenatal stress syndrome: current status. *Psychoneuroendocrinology*, **9**, pp. 3–11.

Ward, I.L. & Ward, O.B. (1985) Sexual behavior differentiation: effects of prenatal manipulations in rats. Handbook of behavioral neurobiology. Vol. 7. Reproduction. In Adler, N., Pfaff, D., Goy, R.W. (eds.). Plenum Press, New York, pp. pp. 77–98.

Weisz, J. (1983) Influence of maternal stress on the developmental pattern of the steroidogenic function in Leydig cells and steroid aromatase activity in the brain of rat fetuses. *Monogr Neural Sci*, **9**, pp. 184–93.

Williams, T.J., Pepitone, M.E., Christensen, S.E., Cooke, B.M., Huberman, A.D., Breedlove, N.J.,

Breedlove, T.J., Jordan, C.L. & Breedlove, S.M. (2000) Finger-length ratios and sexual orientation. *Nature*, **404**, pp. 455–6.

Wisniewski, A.B., Migeon, C.J., Meyer-Bahlburg, H.F., Gearhart, J.P., Berkovitz, G.D., Brown, T.R. & Money, J. (2000) Complete androgen insensitivity syndrome: long-term medical, surgical, and psychosexual outcome. *The Journal of clinical endocrinology and metabolism*, **85**, pp. 2664–9.

Witelson, S.F., Kigar, D.L., Scamvougeras, A., Kideckel, D.M., Buck, B., Stanchev, P.L., Bronskill, M. & Black, S. (2007) Corpus Callosum Anatomy in Right-Handed Homosexual and Heterosexual Men. *Arch Sex Behav*.

Wysocki, C.J. & Preti, G. (2004) Facts, fallacies, fears, and frustrations with human pheromones. *Anat Rec A Discov Mol Cell Evol Biol*, **281**, pp. 1201–1.

Yang, X., Schadt, E.E., Wang, S., Wang, H., Arnold, A.P., Ingram-Drake, L., Drake, T.A. & Lusis, A.J. (2006) Tissue-specific expression and regulation of sexually dimorphic genes in mice. *Genome Res*, **16**, pp. 995–1004.

Zucker, K.J., Bradley, S.J., Oliver, G., Blake, J., Fleming, S. & Hood, J. (1996) Psychosexual development of women with congenital adrenal hyperplasia. *Horm Behav*, **30**, pp. 300–18.

Author Biography

Jacques Balthazart (University of Liège, Belgium) has worked in the field of avian endocrinology since he initiated his thesis work at Liège in the 1970s on the endocrine control of reproductive behavior and reproductive cycles in the Rouen Duck. In his early studies, he was among the group that pioneered the use of radioimmunoassay methods to measure gonadotropins and steroid hormones in the plasma of birds. After a post-doctoral stay at the Institute of Animal Behavior at Rutgers University in the USA, he returned to Liège and established a long-term research program on the neuroendocrine control and sexual differentiation of male-typical reproductive behavior in Japanese quail asking mainly how testosterone is so effective in activating male-typical behavior in one sex but not the other. Overall, he has published more than 400 journal articles and review papers. Nearly all of them are on birds and relate to the action of steroids on brain and behavior. He has organized many conferences on the topic of hormones, brain and behavior, and he has served on the organizing committee of a large number of other meetings. He has and continues to serve on the editorial boards of many journals. He is currently the co-editor of Frontiers in Neuroendocrinology.